Field Guide to Liverwort Genera of Pacific North America

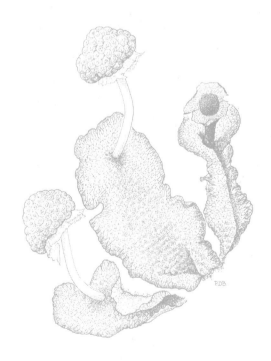

Field Guide to Liverwort Genera of Pacific North America

W. B. SCHOFIELD

Illustrated by
Patricia Drukker-Brammall
and Muriel Pacheco

Global Forest Society
Banff and San Francisco

in association with

University of Washington Press
Seattle and London

To Linda, Muriel, and Pamela

GLOBAL FOREST
Pure Science.

Global Forest (Dr. Reese Halter, founder and president) is dedicated to conserving the world's forests through exploratory research and education. Its mission is to promote conservation of natural resources for future generations, providing vital information to corporate and governmental decision-makers and sharing the wonders of science and nature with children. With the University of Washington Press, Global Forest Society has copublished *Forest Giants of the Pacific Coast*, by Robert Van Pelt, and *Field Guide to Liverwort Genera of Pacific North America*, by W. B. Schofield.

Copyright © 2002 by W. B. Schofield
Printed in the United States of America
Design by Dennis Martin

All rights reserved. No part of this publication may be reproduced or transmitted in any form or by any means, electronic or mechanical, including photocopy, recording, or any information storage or retrieval system, without permission in writing from the publishers.

Global Forest Society
Box 219, Suite 5183, Banff, Alberta T1L 1J7
www.GlobalForestScience.org

University of Washington Press
P.O. Box 50096, Seattle, WA 98145
www.washington.edu/uwpress

This publication was supported by Global Forest Society grant GF-18-2001-176.

Library of Congress
Cataloging-in-Publication Data

Schofield, W. B.
Field guide to liverwort genera of Pacific North America / W. B. Schofield ; illustrated by Patricia Drukker-Brammall and Muriel Pacheco.
p. cm.
Includes bibliographical references (p.).
ISBN 0-295-98194-6 (alk. paper)
1. Liverworts—Pacific Coast (U.S.)—Identification. 2. Liverworts—Pacific Coast (B.C.)—Identification. 3. Hornworts (Bryophytes)—Pacific Coast (U.S.)—Identification. 4. Hornworts (Bryophytes)—Pacific Coast (B.C.)—Identification. I. Title.
QK556.5.P16 S36 2000
588'.3'09795—dc21   2001055654

The paper used in this publication is acid-free and recycled from 10 percent post-consumer and at least 50 percent pre-consumer waste. It meets the minimum requirements of American National Standard for Information Sciences—Permanence of Paper for Printed Library Materials, ANSI Z39.48-1984.

Contents

Preface   vii

Acknowledgments   viii

Introduction   1

Structure of Liverworts and Hornworts   3

Liverworts and Hornworts Compared to Mosses   7

How to Collect Liverworts and Hornworts   7

History of Collectors   13

Habitats of Liverworts and Hornworts   16

Seasonality of Liverworts and Hornworts   18

Distribution Patterns of Liverworts and Hornworts in the Region   19

Critical Determination   20

Latin Names   21

Implications of Liverworts to People   21

Key to Determine Genera of Liverworts and Hornworts in the Region   23

Use of the Hand Lens   25

Use of the Key   25

Genera of Liverworts and Hornworts   35

Glossary   213

Checklist of Liverworts, Hornworts, and *Takakia*   217

References   223

Index   225

Preface

For most people, the liverworts and hornworts are probably the least known of the green land plants. This is unfortunate, because they exhibit great beauty and show remarkable fitness to their environment. Many naturalists confuse these plants with mosses or lichens and have vague ideas concerning the great diversity of their form and ecology. To recognize most species, microscopic details are necessary, although once species are learned, it is possible to recognize many without microscopic examination. Most genera, on the other hand, can be identified using features visible with the naked eye or with a 12x hand lens. Few attempts have been made, however, to produce a guide that is based on such readily visible features.

In a flora with a relatively limited number of genera, however, the creation of such a guide is possible. The guide presented here undoubtedly has shortcomings, but the beautiful illustrations of Patricia Drukker-Brammall and Muriel Pacheco present a very good idea of the elegance as well as the features that distinguish the genera. The written comments present other characters valuable for identification.

Many genera are represented in our flora by a single species, and this makes identification easier because genera are separated based on major distinguishing features. Where only a single species is illustrated, a very common one is chosen that represents the main features of the genus. Some of the rarer species and even many common ones in the region are impossible to name with confidence with this book alone. In most cases these represent divergent elements in the genus, and their determination even to genus requires microscopic examination.

This book attempts to present these plants as organisms of great inherent fascination, but neglect has inhibited the accumulation of detailed information on their biology and distribution. Knowledgeable amateurs can make valuable contributions to broaden this understanding.

I have included the genus *Takakia*, which is neither a liverwort nor a hornwort. It is of such inherent interest, however, that it has been added. It has superficial resemblance to a liverwort and therefore may be mistaken for one.

W. B. S.
October 2001

Acknowledgments

My greatest debt is to Dr. Reese Halter and Global Forest; they have underwritten the publication of this book. I am immensely grateful.

Most of the illustrations were financed by the Dean's Research Fund of the University of British Columbia as well as by a bursary from the Killam Foundation. Most of the field research and collecting on which familiarity with the living plants has been gained was supported by the Natural Sciences and Engineering Research Council of Canada. Marion Crosson, Judy Heyes, Tami Chappell, Lana Chan, Lebby Balakshin, and Regina Yung have typed the manuscript from a draft of often baffling illegibility. Judy Godfrey, Rudy Schuster, Sinske Hattori, and Hiroshi Inoue have greatly enriched my knowledge of liverworts. Alan Whittemore generously gave access to his unpublished treatment of California's liverworts and hornworts. I am deeply indebted for this assistance. For comments on the manuscript, I am indebted to Judy and Geoff Godfrey, although I may not have followed their good advice at all times; any errors that have slipped by are my regretful responsibility.

I again acknowledge my great debt to the talented illustrators: Patricia Drukker-Brammall and Muriel Pacheco. Their beautiful drawings are vital as well as richly decorative. Adolf and Oluna Ceska and David Wagner have generously provided specimens for illustration. Shona Ellis tested the key and revealed problems, some of which are solved. To all of these colleagues, I am most grateful. Patrick Williston has been a friendly presence who has assisted in the completion and publication of this book.

I want to acknowledge the support of the University of Washington Press, in particular thanks to Pat Soden, who sponsored the project; Mary Ribesky, who edited the manuscript; and Dennis Martin, who designed the book.

Introduction

Liverworts and hornworts, although widespread and often abundant in western North America, tend to be less bountiful than mosses. Near the coast, especially in the wetter climates from northern California to Alaska, they are often luxuriant. They form mats and tufts on cliffs, near watercourses and ponds, and in humid seepy areas. Even in open forest, liverworts sometimes abound on trunks and branches of trees and shrubs, and form extensive carpets or turfs on boulders and in somewhat shaded moist areas, especially on rotten logs and stumps. In peatland and swamps, liverworts are sometimes abundant. Some liverworts are found in gardens, especially in sites that are shaded and moist and where other plants cannot thrive, but they are neither common nor very conspicuous. Some liverworts are confined to soil that is seasonally moist, and they vanish with the dry seasons; a few are strictly aquatic. In California and southern Oregon, hornworts are frequent in humid, somewhat shaded areas, but are evident only in the spring.

Their seeming infrequency adds to the fascination that these plants offer. Enjoyment of their beauty of form is also of considerable attraction. Discovery of the rarer species and study of their biology and distribution offer special rewards. It is possible to discover an entirely new species in the western North American liverwort flora, but it usually involves a far more intensive knowledge than this book provides. Indeed, a person should be discouraged from describing new species without consultation with specialists. The gaining of expertise results from an insatiable interest in the living plants, from their very careful study, and from a thorough knowledge of the pertinent literature.

Structure of Liverworts and Hornworts

Liverworts show great variety in form and can be confused by the novice with other organisms. Thallose forms (i.e., those in which the form of the plant is somewhat straplike) are sometimes confused with lichens. In liverworts, however, thalli are usually green whether living or dead, although some turn black when dead; the same is true for hornworts, in which the plant body is always thallose in our local genera. Some thallose lichens superficially resemble liverworts. In lichen genera, however, since most of the organism is usually constructed of a fungus, cellular pattern is not discernible on the surface, and the green color is often obscured in the dried state because the algal layer is generally located just beneath the cortex. Translucent when moist, the cortex becomes opaque when dry and obscures the green color.

The thallose growth form of the conspicuous liverworts provides the popular name of this group of plants. To early herbalists, some thallose genera suggested resemblance to a liver, and the herbalists interpreted this as a divine message that marked this plant as an effective cure for liver ailments, hence "liver plant." Most liverworts, however, do not produce flattened thalli. Indeed, most are mosslike in appearance. This led to the name "scale mosses," but the term has never achieved wide popular usage.

Liverworts are classified with the major plant group termed *bryophytes*. This major evolutionary line includes all plants in which the dominant and long persistent stage of the life cycle is the sexual stage. This stage, the *gametophyte*, as the term indicates, bears the sex organs that, in turn, produce the gametes (i.e., sperms and eggs). The gametophytic stage is photosynthetic and generates most of the nutrient for the partially parasitic stage that forms the spores, the *sporophyte*. Spores serve as an important means of spreading a species from one site to another, a process that usually involves air masses that move near the surface of the ground.

The male sex organ is the *antheridium*, a multicellular sac that produces many sperms; the sperms are released when an antheridium is mature and is bathed in water. The cells of the mature antheridium take up water until they become very swollen. Each of the many cells within the antheridium contains a single sperm, while the multicellular jacket of the antheridium simply forms the sac that encloses these sperm-producing cells. Cell walls that enclose the sperms dissolve, the antheridium jacket is ruptured, and the sperms are released into the water. The sperms, with fatty material among them, quickly move to the surface-film of the water, and instantly spread, as a mass, over the surface. Each sperm has two hairlike threads (or *flagella*) and the undulation of these causes the sperm to swim in a corkscrew path in the water.

The female sex organ, the *archegonium*, when mature, is bottle-shaped and encloses a single egg. The neck of this bottle is composed of a jacket of cells that encloses a single row of cells to form the neck canal that leads to the egg within the bottle. When the mature archegonium is bathed in water, the walls of these neck canal cells dissolve and release the sugary cell contents into the surrounding water. This sugary material moves out of the neck canal and spreads outward into the watery medium. When a sperm contacts any dilution of this substance, it swims toward the area of the greatest concentration, thus down the neck canal to the egg. When the sperm burrows through the wall of the egg that remains within the archegonium, its nucleus unites with the egg nucleus, and the resulting cell possesses a nucleus that combines the nuclear material of both sperm and egg. This is the *zygote*, the first cell of the sporophyte. This zygote undergoes many cell divisions within the archegonium to produce the embryonic sporophyte. While the embryo develops, it is protected by a sheath of tissue that has formed from the growth of the archegonium wall and associated cells; this is the *calyptra*. In all bryophytes the sporophyte is intimately attached to the gametophyte, with the foot pushing itself among the cells of the gametophyte and serving as an active area of transfer of water, minerals, and nutrients from the gametophyte to the developing sporophyte.

In all bryophytes the sporophyte produces a single sporangium. In liverworts this usually terminates a stalk (*seta*) that pushes the mature sporangium above the gametophyte and improves opportunities for spore dispersal by moving air currents. In liverworts, the seta is usually a very fragile structure and elongates rapidly as soon as the sporangium is mature. The elongating seta pushes the sporangium through the ruptured calyptra, leaving the calyptra attached at the seta base. The seta tends to be colorless and remains erect through turgor pressure within the cells until its cells dry out, when it collapses. In most cases the sporangium opens by longitudinal lines as it dries out, exposing its contents: springlike cells (*elaters*) and spores. These elaters uncoil abruptly and jump into the air, breaking up the spore mass and throwing spores into the air currents that caused the drying. Spores, when they fall in a site with suitable conditions of moisture and light for germination, ultimately produce new gametophytes.

Bryophytes also reproduce by production of vegetative structures termed *gemmae*. In leafy liverworts these are produced on leaf surfaces and margins, or completely replace leaves. Each gemma, generally as small as a dust grain, consists of one or more cells. It is readily wind-borne to a distant site where it germinates much like a spore and can produce a new gametophyte. In a few thallose liverworts, gemmae are produced in gemma cups, as in *Marchantia* and *Lunularia*. Water drops that splash into the cup throw the gemmae away from the thallus to a new site. Some thallose liverworts produce gemmae on the thallus margin (e.g., *Metzgeria*), others in flasks (*Blasia*) or within cells, especially near the thallus apex (*Riccardia*). In all bryophytes any fragmentation of a living gametophyte can result in each isolated living fragment growing to form an independent gametophyte.

In our local hornworts the gametophyte is always a green thallus, usually irregularly circular in outline with narrow or broad lobes radiating outward from the center of the thallus. The sporophyte is, when young, a green tapered cylinder (the "horn" of the hornwort); it differs from other bryophytes in its continuous growth throughout the growing season. As the apex of the sporangium ripens, the lower portion remains green. The apex of the sporangium opens by two lines that extend downward toward the base as ripening continues and elongation of the sporangium pushes the ripened apical region upward. Elaters are among the spores, and these aid in scattering the ripened spores to the air currents that cause them to dry.

In some hornworts the thallus survives the unfavorable season by the isolation of lobe tips through decay of the center of the thallus. Each of these lobe tips contains an apical cell protected by dead remains of other cells as well as nutrient-rich cells. When favorable conditions return, each of these tips (termed *tubers*) can germinate to produce a new thallus.

Thallose liverworts show considerable variety in form. The structural simplicity in *Pellia* presents a thallus in which the central portion is many cells thick, the uppermost cells are chlorophyll-rich, and the lowermost cells are mainly colorless. This thallus also shows very simple branching in the broad lobes. Colorless (or sometimes colored) *rhizoids* attach the thallus to the substratum; these are mainly on the thickened central portion of the thallus. In this type of thallus the male sex organs are embedded in tiny wartlike protuberances on the upper surface of the thallus. These open by a pore that enlarges as the thallus ages and the single antheridium within matures. Several archegonia are enclosed in a sleevelike tube (*involucre*) on the upper surface of the thallus and near the tip of a lobe.

A somewhat more specialized thallus is represented by the thallus of *Metzgeria*, where there is a stemlike cylindrical portion from which rhizoids emerge on the undersurface. On both lateral surfaces of this stem extend a band of tissue several cells wide and one cell in thickness. All cells contain chlorophyll, except the rhizoids. Antheridia are enclosed in hemispherical branches on the undersurface of the stem, while archegonia are enclosed within a cylindric involucre, also on the undersurface of the stem. These specialized sexual branches are usually covered in spinelike, short hairs.

Even more specialized is the thallus of *Apotreubia*, in which there are very regular leaflike lobes longitudinally attached on the lateral sides of the stemlike central portion. On the upper surface of the stem, usually associated with each lateral lobe, is a leaflike scale. Most cells of the upper surface of this thallus are chlorophyll-rich, while scattered cells contain single complex oil bodies. The thallus is attached to the substratum by rhizoids. Antheridia and archegonia are exposed or associated with the dorsal scales.

The most complex thalli are represented by *Marchantia* and *Conocephalum*. The thallus contains a single layer of dorsal chambers, each with a single dorsal pore and containing chlorophyll-rich filaments. On the undersurface of the thallus, rhizoids are abundant, forming a cottony mass of threads that are parallel with the thallus length and other threads that attach the thallus to the substratum. Also on the undersurface of the thallus are scales, often with a purplish pigment. These are most obvious near the apices of the lobes and protect the apical cells. In these genera, too, the sex organs are borne on specialized receptacles. In both genera, the archegonia are on an umbrellalike receptacle that bears the sporophytes on its undersurface. As it ages, the receptacle elongates a stalk that carries the developing sporophytes upward above the thallus surface and exposes the ripe sporangia to moving air currents that carry the spores away. In *Marchantia*, too, the antheridia are also borne on a stalked receptacle. In *Conocephalum*, however, the antheridial receptacle forms a circular pad near the thallus margin.

Other thalli, like those of *Ricciocarpos*, are composed mainly of air chambers, while still others have very narrow vertical chambers, as in some species of *Riccia*. These thalli have internal sex organs and sporangia.

In texture, thalli vary considerably, from soft and flexible to somewhat rigid. Sometimes the thallus is succulent, and when dead and dry becomes somewhat papery. In others the thallus changes little in consistency and color from the living to the dead condition. In some, the thallus turns black when dead.

Most thalli are green when living, although the scales and margins of some are a deep wine-red to purple or colorless. The thallus of local hornworts is dark green when living, usually has many narrow lobes radiating outward from the center, and is somewhat succulent. When dead, the thallus of hornworts usually turns black.

Leafy liverworts show considerable diversity in general form. In *Haplomitrium*, for example, the colorless creeping stems, the lack of rhizoids, and the radially arranged leaves make this genus highly distinctive. In most other leafy liverworts, the leaves are in either three or two rows. Rhizoids are confined to the undersurface of the stem and are generally colorless, but in some species of *Jungermannia*, for example, rhizoids are purplish. In *Takakia* rhizoids are absent, but a colorless rootlike system is often produced.

Leafy liverworts in which leaves are in three rows usually have two rows of lateral leaves and a single row of underleaves. Generally the lateral leaves are attached at an oblique angle, thus the leafy shoots often appear somewhat dorsiventrally flattened. The lobing of the leaves as well as the manner in which the leaves overlap and the margins curve give the leafy shoot a distinctive appearance.

While thalli generally recline, many leafy hepatics have the shoots erect or suberect, and form tufts or interwoven mats. Those that recline are sometimes firmly affixed to the substratum or are loosely attached. In many, the leafy shoots are intermixed among other plants, especially mosses.

Color of leafy liverworts varies. Most are some shade of green, from a vivid dark green to pale whitish or bluish green or yellow green. In these the color is sometimes altered from the moist to the dried state. Other leafy liverworts are yellow to golden, rich orange red, wine red, dark purple, black, or brown. Brown liverworts can be a pale rusty brown, dark brown, or golden brown. In most liverworts the pigment is in the cell walls, but in some the color results from pigments in oil bodies within the cells.

This diversity of color, texture, and growth form makes the leafy liverworts highly attractive plants that greatly enhance the beauty of the vegetation.

Liverworts and Hornworts Compared to Mosses

Liverworts and hornworts differ from mosses in many features. No mosses are thallose, whereas many liverworts and most hornworts are thallose. Leafy liverworts differ from mosses in several features easily viewed by a hand lens and the naked eye. Microscopic features are also significant.

The list below compares leafy liverworts to mosses, with liverworts noted first, then contrasted with mosses. Microscopic features are not included. Although not applicable to *all* mosses and leafy liverworts, these features are usually reliable.

How to Collect Liverworts and Hornworts

Some liverworts and hornworts are seasonal in the appearance of a gametophyte visible to the naked eye. Although most liverworts are perennial, during the dry season some dry up and become dormant and rather nondescript in appearance. For most liverworts and all hornworts, early in the growing season is the best time for observation. At low elevations the best season is the spring or autumn, whereas at higher elevations early summer to autumn are the most favorable seasons. At these times, too, sporophytes are more frequently present, and contribute information

	LIVERWORTS	MOSSES
1	Leaves in 2 or 3 rows	Leaves usually in more than 3 rows on stem
2	Leaves often lobed	Leaves not lobed
3	Leaves without midrib	Leaves often with midrib
4	Rhizoids unbranched	Rhizoids branched
5	Rhizoids usually colorless	Rhizoids usually reddish brown
6	Perianth sleeve usually around archegonia	Perianth absent
7	Seta usually colorless	Seta usually pigmented (brownish or yellowish when mature)
8	Seta collapsing when spores are shed	Seta persistent after spores are shed
9	Seta fragile	Seta wiry
10	Sporangium maturing before seta elongates	Sporangium formed after seta elongates
11	Sporangium usually opening by longitudinal lines	Sporangium opening by an apical lid (operculum)
12	Sporangium usually containing elongate elaters and spores	Sporangium containing only spores
13	Sporangium lacking teeth around the mouth	Sporangium usually with teeth around the mouth
14	Calyptra left at base of seta when sporangium raised above the gametophyte	Calyptra forming an apical cap on the sporangium at the tip of the elongate seta

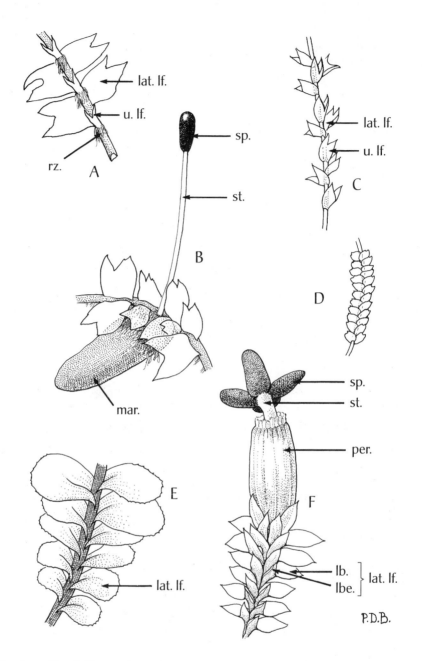

Fig. 1. Structure of Leafy Liverworts

(A) *Lophocolea cuspidata*, undersurface showing lateral leaf (lat. lf.), underleaf (u. lf.), and rhizoids (rz.); (B) *Geocalyx graveolens*, upper surface of plant with sporophyte emerging from the subterranean marsupium (mar.), or pouch; the sporophyte is composed of a seta (st.) and sporangium (sp.); (C) *Hygrobiella laxifolia*, showing the similarity of underleaf and lateral leaf.; (D) *Bazzania denudata*, showing the overlapping leaves (incubous); (E) *Plagiochila porelloides*, showing the leaves that underlie each other (succubous); (F) *Douinia ovata*, showing the sporophyte (opened) emerging from the perianth (per.) and the lateral leaf with unequally sized lobe (lb.) and lobule (lbe.).

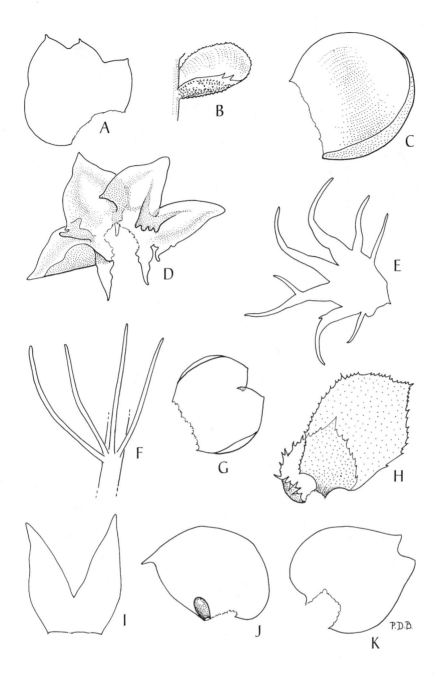

Fig. 2. Diversity of Leaf Form in Liverworts

(A) *Tritomaria* with three broad lobes of unequal size; (B) *Cololejeunea* with small, toothed lobule and lobe with marginal small teeth; (C) *Odontoschisma*, unlobed leaf; (D) *Tetralophozia* with four equal lobes; (E) *Ptilidium* with many lobes; (F) *Blepharostoma* with four filamentous lobes; (G) *Marsupella*, equally bilobed; (H) *Scapania*, unequally bilobed, with lobule uppermost; (I) *Anthelia*, deeply and equally bilobed; (J) *Frullania*, unequally bilobed, with very specialized helmet-shaped lobule and pointed lobe; (K) *Anastrepta* with shallow lobes at leaf tip.

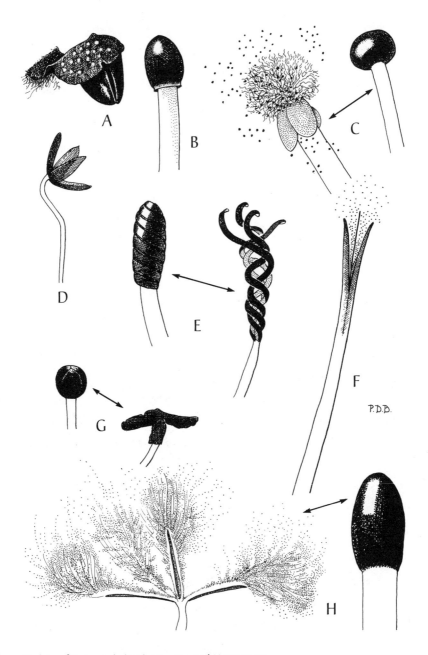

Fig. 3. Variety of Sporangia in Liverworts and Hornworts

(A) *Targionia*, showing ventral sporangium; (B) *Blasia*, elongate sporangium; (C) *Pellia*, opening sporangium (left) with mass of elaters, and unopened sporangium (right); (D) *Diplophyllum*, sporangium opened to show four longitudinal divisions; (E) sporangium of *Gyrothyra* with helical lines of opening, unopened sporangium on left and opened sporangium on right; (F) *Anthoceros*, sporangium splitting from apex downward as it elongates; (G) *Gymnomitrion*, unopened sporangium on left; (H) *Riccardia*, opened sporangium on left, forcibly ejecting spores from pulsating elaters.

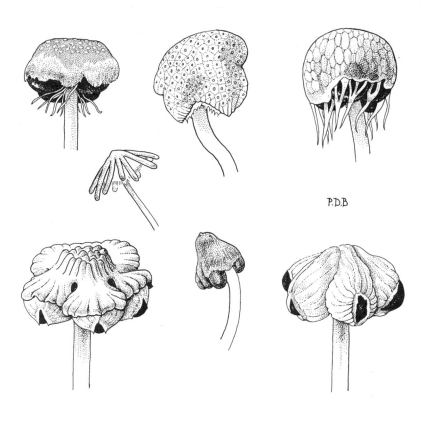

Fig. 4. Receptacles of Thallose Liverworts

Top row (left to right): *Mannia*; *Preissia*; *Asterella*

Middle row: *Marchantia*

Bottom row (left to right): *Peltolepis*; *Conocephalum*; *Sauteria*

that is useful in identifying genera. Some liverwort genera, however, do not produce sporophytes in the area treated here, or sporophytes are rare. In consequence, material lacking sporophytes should be collected when none can be found with sporophytes.

These plants are easy to collect and preserve, although some change their form from moist to the dried condition, making any future identifications more difficult. Hornworts and some thallose liverworts, for example, turn from green to black in the dried condition, and even when thoroughly rewetted, do not regain their living appearance. Most thallose liverworts shrink considerably when drying, and thus lose much of their living character. For leafy liverworts, and some thallose genera, color and form alter only slightly when drying, and these plants regain their living appearance when moistened.

Some liverworts adhere very firmly to their substratum. To remove them often requires carefully teasing them away with a pocketknife or chisel. This should be done so that the branching pattern is not destroyed or the material greatly fragmented. Sometimes, when on wood or bark, it is necessary to cut away the substratum with the attached plants.

For hornworts and liverworts that grow on soil, it is usually necessary to remove some of the soil with the plant. Loose soil should be cleaned away carefully so that it does not damage the dried specimen in its packet.

Enough material should be collected in order that some can be retained for reference purposes. When further specimens are collected and identified, these can be compared with earlier collections. In time, even when the species cannot be determined by using the keys provided here, a broader idea can be gained concerning the diversity within a genus in the area. Added to this, the specimens can be shown to an expert, or another student, and the probability is increased that a more precise determination can be made.

Specimens can be air-dried without pressing, in a paper packet. It is important to tease out the material so that it fits into an envelope-like packet without producing an unwieldy lump. The material should show, as much as possible, its living appearance. Temporary packets can be made from newspaper cut into the dimensions of typewriter paper sheets (approximately 8 x 11 in. or 27.5 x 21 cm). This sheet is folded upward with a flap about one third the length of the page, then the two ends are folded inward to form an envelope. The rest of the length of the sheet is folded downward to form a flap over the closed packet. Material sufficient to fill this packet should be collected, if possible.

Information should be recorded when the collection is made:

1. Habitat (e.g., rock, rotten wood, etc.). If on a tree, it is important to note the kind of tree (maple, cedar, etc.). If the type of rock can be identified, this information is often instructive (granite, limestone, etc.). A note on moisture (wet, submerged, dry) provides more useful information about the habitat. Nature of the surrounding vegetation is sometimes helpful to know (tundra, pine forest, grassland, etc.). Elevation of the collection locality is very significant as is the exposure of the habitat (sunny, shaded).

2. Collector's name.

3. Collecting locality (e.g., Takakia Lake, Moresby Island, Queen Charlotte Islands, B.C., 52° 56′N, 132° 04′W).

4. Date collected (e.g., July 19, 1927).

5. When the specimen is identified, the Latin name should be recorded on the packet.

When the specimen is dry, it can be transferred to a more permanent packet made of typewriter paper. The label can be placed on the flap of this packet.

Storage of specimens can be accommodated in shoe boxes, with packets filed like index cards. A stiff card beneath the specimen in the packet helps protect the dried material.

It is useful to keep a notebook record of collections and to give a collecting number to each specimen. This provides a record of the material from each collecting locality and gives a collector an idea of the botanical richness of that locality as well as an impetus to explore the site thoroughly for all the diversity it can provide.

An amateur can assemble a valuable hepatic collection that presents a useful documentation of the flora of a given area by ensuring that collections are well recorded and carefully prepared. Such a collection can be presented to a permanent collection (e.g., university, state, or provincial herbaria). Herbarium collections are invaluable as a permanent documentation of species from an area and also serve as a reference collection to clarify the identity of species when compared with other collections.

A cautionary note on collecting: a collector should collect material carefully, and be sure that the species is not destroyed in its habitat. This is achieved by removing only a portion of the population, leaving the remainder undisturbed. Bryophytes are as vulnerable to extinction as are other organisms, and are important members of the plant community.

History of Collectors

Alaska

The first collection of liverworts from Alaska appears to have been made casually by J. T. Rothrock, who reported six species in 1867. Indeed, most collections of liverworts have been taken incidental to other research. The main researchers who have assessed the liverwort flora have not made the collections, but have determined material collected casually by others. Alexander Evans and Herman Persson, the eminent Swedish bryologist, are especially important researchers. T. C. Frye made major collections in southeast Alaska in 1913. Other early collectors include Grace Cooley (1891), J. M. Macoun (1891), R. S. Williams (1898–99), T. H. Kearney and J. V. Coville (1899), William A. Setchell (1899), William Trelease (1899), A. B. Foster (1913), and G. B. Rigg (1913). W. C. Steere was the first bryologist to make major collections of liverworts in the arctic portions of the state, and he concentrated his efforts mainly on the arctic. Steere's extensive bryological experience and knowledge, coupled with an extraordinarily discriminating eye, enabled him to collect with considerable ability. This, with his inherent fascination for living plants, enabled him to discover many intriguing species. He also attracted the interest of other bryologists to the same region and built up an exceptionally sound foundation of knowledge, which he summarized in 1978. Steere discovered many hepatics that proved to be either novelties or new to the flora of North America. Among these are the genera *Mesoptychia*, *Ascidiota*, and *Calycularia*.

Other recent collectors who have made important collections of liverworts are: M. Barker, P. G. Davison, W. J. Eyerdam, K. Holmen, D. H. Horton, H. Inoue, Z. Iwatsuki, N. Konstantinova, S. Looman Talbot, H. A. Miller, B. Murray, A. Potemkin, W. B. Schofield, R. M. Schuster, H. Shacklette, A. J. Sharp, D. Smith, S. Talbot, and I. A. Worley.

It is of interest that Herman Persson first gained the basis for his understanding of the liverwort flora of Alaska through study of material that he teased out from collected specimens of vascular plants. These scraps of material yielded a surprisingly comprehensive list of the basic hepatic flora of the state, but revealed few of the hepatics that consistently occupy sites poor in vascular plants. Although I. A. Worley's collections from the Alexander Archipelago and the collections of P. G. Davison, W. B. Schofield, A. J. Sharp, D. Smith, and S. Talbot from the Aleutian Chain have given a base upon which to build, the details of that region require concentrated fieldwork by an experienced field hepaticologist.

At a very conservative estimate, less than 5 percent of the territory within Alaska has been explored adequately for hepatics.

British Columbia

The first major collector of liverworts in British Columbia was the naturalist to the Geological Survey of Canada, John Macoun. His collections of all groups of organisms provided the basis for the first assessment of the biological diversity in Canada. He made the first collection, among many other liverworts, of the type specimen of *Cololejeunea macounii*. Although collected at Hastings Mill, now part of Vancouver, the species has not been found in the Vancouver area since that time. Regrettably, Macoun was somewhat casual in the labeling and distribution of his collections, thus a full set of his specimens is not found at the National Herbarium of Canada, but is scattered in many European and North American herbaria.

A. H. Brinkman, a remarkably competent amateur, made a number of important collections of liverworts in eastern British Columbia, among them the type specimen of *Cephaloziella brinkmanii*. The species has never been re-collected.

Sarah Gibbs made several important collections of liverworts from the west coast of Vancouver Island during the time when the University of Minnesota had a biological station at Port Renfrew. Her collections yielded the recognition of a new species: *Odontoschisma gibbsiae*.

T. C. Frye and S. Flowers made small collections in the province, mainly on Vancouver Island in the mid-twentieth century. In 1956, Herman Persson made an important collection of liverworts from the Queen Charlotte Islands. Among his collections was the unusual plant *Takakia lepidozioides*, previously collected from Japan. His collections from the Queen Charlotte Islands hinted at the bryological richness and importance of this region.

T. Ahti and L. Hämet-Ahti made collections in Wells Gray Park that remain unreported.

W. S. Hong and D. H. Horton have made important recent collections of liverworts in British Columbia. Dr. Hong's collections have contributed to his floristic studies of the liverworts of western North America while Dr. Horton's collections have emphasized phytogeographically significant species.

The most important recent collectors of liverworts in the province are Judith and Geoffrey Godfrey. Their collections form the basis of the first intensive study of liverworts and hornworts of British Columbia. Judith Godfrey's researches led to the recognition of two new species: *Frullania hattoriana* and *Jungermannia schusterana*, an unusual subspecies of a species otherwise known from China, *Scapania hians* ssp. *salishensis*; and the recognition of a distinctive endemic western North American genus, *Schofieldia*, with a single species *S. monticola*.

R. M. Schuster has made a number of collections of liverworts from the province, among them a collection from the Queen Charlotte Islands that led to the recognition of a new genus: *Dendrobazzania*.

The most extensive collections of liverworts and hornworts for the province are those of W. B. Schofield. These collections, made throughout the province, have been accumulated from 1960 until the present and have resulted in the discovery of numerous taxa previously unknown from North America. Among the genera are *Apotreubia, Chandonanthus, Dendrobazzania, Mastigophora,* and *Schofieldia*.

Washington

Although T. C. Frye was an active hepaticologist through his long life in Washington State, he devoted very little time to comprehensive fieldwork in the state. As a possible consequence of Frye's long residence, there has grown the assumption that the liverwort flora of the state is well documented; such is not the case. It is probable that the most comprehensive collections made in the state are those of Judith and Geoffrey Godfrey, J. Harpel, W. S. Hong, and W. B. Schofield. The northwestern portion of the state has been reasonably completely documented, but much of the rest of the state is represented by a few small collections made incidental to other work.

Clark and Frye's *The Liverworts of the Northwest*, published in 1928, represented an assessment of the liverwort flora as understood at that time. Unfortunately this work was based essentially on published records and the fieldwork supporting it was extremely restricted. They state that the treatment was done "usually without the material at hand for study." Although it is probable that most liverworts likely to be found in the state have been documented by specimens, the knowledge of distribution is based on very limited material, and comprehensive fieldwork is greatly needed.

Oregon

The first attempt to assess the liverwort flora of Oregon was made by W. S. Hong in 1978. As his report demonstrates, no concerted attempt has been made by a hepaticologist to document the flora thoroughly. The first collector appears to have been J. Scouler in 1825. Recent students who have accumulated important collections are J. Christy, Judith and Geoffrey Godfrey, W. S. Hong, H. A. Miller, W. B. Schofield, and D. Wagner. E. L. Sanborn, in 1929, produced a manual to liverworts and hornworts of western Oregon. Besides her own important collections, she mentions major collections made by A. S. Foster, T. C. Frye, and Lois Clark. As is true of the state of Washington, it seems possible that a broad understanding of the liverwort flora in Oregon is reasonably complete, but detailed knowledge of distribution patterns and knowledge of the rarer species is incomplete.

California

The first comprehensive liverwort flora published for the Pacific Coast is that of M. A. Howe (1899) for California. This is a thoroughly scholarly work and remains valuable today in spite of the outdated nomenclature. Its great value lies in the careful study of living material and the familiarity with the plants in nature.

The first collections of liverworts made in the state were those of Archibald Menzies, taken in the latter part of the eighteenth century. The first important collector was H. N. Bolander, who collected many specimens while a resident in California from 1861 to 1878. He was a fine collector with a keen eye, and is commemorated by several species of liverworts named in his honor, including *Scapania bolanderi, Frullania bolanderi, Asterella bolanderi, Marsupella bolanderi, Porella bolanderi,* and *Radula bolanderi*. The collections of M. A. Howe made in the latter part of the nineteenth century are also very important. He described the unusual genus *Gyrothyra* based on these collections. In the early part of the twentieth century, C. C. Kingsman reported on the hepatics of southern California. In recent years, major collections of liverworts have been made by W. T. Doyle, D. Wagner, W. B. Schofield, D. L. Sutcliffe, and A. T. Whittemore, and incidental, but important, collections have been made by P. Yurky and D. H. Norris. A. T. Whittemore has a flora in manuscript (1999); this should stimulate a renewal of research. Liverworts from mountainous regions of California are poorly documented, and collecting in late winter and spring in much of the Central Valley and the near-coastal areas is likely to yield a number of unexpected genera and species.

As is clear from the history of collections, much remains to be done to establish an adequate knowledge of the liverwort and hornwort flora of the region. Interested and dedicated amateur botanists resident in the area can make invaluable contributions. Intensive collections made throughout many growing seasons in a specific area can yield important records of uncommon species and provide new information concerning seemingly well-understood species.

It should be emphasized, however, that collections should not be made indifferently, especially when species have been noted to be rare. Liverworts and hornworts are sometimes very discriminating in the habitats they can occupy. Carelessness can destroy a local population or destroy the suitable habitat and lead to the extinction of that species in its local area.

Habitats of Liverworts and Hornworts

The structure of the gametophyte determines, at least in part, the habitat of many liverworts and hornworts. Thallose genera generally grow on soil, although a few occur on rock, trees, or shrubs, and some occur in water. Thallose genera that grow in sites that dry out frequently tend to be small in size or confine their growing season to a limited interval of time. In Pacific North America, *Metzgeria* and *Apometzgeria* are commonly epiphytes or occur on rock faces. Other thallose genera occur on seasonally dry soil; examples are *Anthoceros, Aspiromitus, Asterella, Cryptomitrium, Fossombronia, Geothallus, Mannia, Riccia, Sphaerocarpos,* and *Targionia*. Some of these genera are restricted to the drier climates in the southernmost part of British Columbia and extend southward to southern California (*Anthoceros, Aspiromitus, Sphaerocarpos,* and *Targionia*); all of these genera produce sporophytes in the spring. Others are predominantly Californian (*Cryptomitrium, Geothallus*).

Thallose genera that occur on soil in somewhat shaded and humid sites include *Apotreubia, Athalamia, Blasia, Bucegia, Calycularia, Conocephalum, Lunularia, Marchantia, Peltolepis, Preissia, Reboulia, Riccardia,* and *Sauteria*. These are found on shaded banks, on cliff shelves and bases, and among boulders; a few are in gardens. Thallose genera with species that grow in wet sites include *Aneura, Anthoceros, Conocephalum, Fossombronia, Marchantia, Moerckia, Pellia, Riccardia, Riccia,* and *Ricciocarpos*. Indeed, *Ricciocarpos* is a floating aquatic, while *Riccia fluitans* is suspended under the water surface. The genus *Riella* is a submerged aquatic on mud. At higher elevations *Aspiromitus* occurs in damp sites. *Marchantia* has wide moisture tolerances; it occurs in wet sites in subalpine and alpine elevations, forming extensive mats among mosses colonizing snow-melt stream margins, but is found also near sea level, usually in somewhat shaded sites. *Moerckia* and *Aneura* are frequent on peaty banks and depressions where snow or water persists for long periods. *Fossombronia foveolata* occurs in similar sites, especially on margins of lakes and ponds. *Riccardia multifida* and *R. chamedryfolia* frequently occur in wet sites, with *R. multifida* frequent in swampy depressions in woodland and on dripping cliffs. *Calycularia* is often in shaded humid grottoes.

The leafy liverworts show a vast diversity of habitat tolerances. In some genera all species grow in a similar habitat, while in others, each species differs markedly in its ecology.

Aquatic Habitats

Few leafy liverworts are submerged through their entire growth period. Some, however, do grow in water, usually attached to a solid substratum. *Anthelia, *Chiloscyphus, *Cladopodiella, *Gymnocolea, Hygrobiella, *Jungermannia, *Marsupella, Mylia, Nardia, Pleurocladula, Pleurozia, *Scapania,* and *Schofieldia* all have species that grow in wet habitats; those marked * have species that are submerged or flushed with water through much of their life.

Wetlands

Some of the same genera of aquatic sites have species that occur in the vegetation mat of wetlands or boggy areas. Indeed, other genera also have species in these damp sites: *Calypogeia, Cephalozia, Chiloscyphus, Diplophyllum, Kurzia, Lophozia, Mylia, Odontoschisma, Plagiochila, Scapania, Tritomaria.* The specificity of some of these genera to sites that are wet for a very circumscribed annual period is very marked. For example, *Anthelia* generally occupies sites that are very wet (sometimes submerged) for all but two or three months, while some populations of *Chiloscyphus* and *Scapania* are wet throughout the year.

Cliff and Boulder Faces

A number of genera have species that grow firmly affixed to perpendicular or horizontal rock surfaces. Some grow mainly on rock that is rich in silica and therefore often somewhat acid in chemical reaction. Others are most frequent on rock rich in lime, thus alkaline in chemical reaction. Most, but not all, species tolerate some extended dry periods.

(a) Siliceous Rock: *Cephaloziella, Diplophyllum, Douinia, Frullania, Gymnomitrion, Herbertus, Jungermannia, Marsupella, Plagiochila, Porella, Radula, Scapania, Sphenolobopsis, Takakia, Tetralophozia,* and *Tritomaria.*

(b) Calcareous Rock: Some species of *Jungermannia, Lophozia,* and *Scapania* occur predominantly on calcareous rock.

Epiphytes

Epiphytes are plants that grow on the surface of other plants. It is in the humid coastal forest that epiphytes abound, from northern California to southeastern Alaska. Indeed, in this region some genera that are otherwise predominantly terrestrial become epiphytic on tree trunks and branches. The following genera have epiphytic species: *Anastrophyllum, Barbilophozia, Bazzania, Blepharostoma, Calypogeia, Cephalozia, Cephaloziella, Cololejeunea, Diplophyllum, Douinia, Frullania, Geocalyx, Herbertus, Jamesoniella, Jungermannia, Kurzia, Lepidozia, Lophocolea, Lophozia, Mylia, Odontoschisma, Plagiochila, Porella, Ptilidium, Radula, Scapania,* and *Tetralophozia.*

Rotten Logs or Stumps/ Usually in Forest

In forest, wood in various stages of decomposition harbors a number of liverworts that sometimes form dense turfs or mats for a limited period as decomposition of the wood progresses. In early stages the persistent bark has several genera; as the bark decomposes, other genera replace the early colonists. Finally the wood becomes part of the substratum of the forest soil and a different spectrum of liverworts enters. In forests where moisture is ample, either through dissection by watercourses or through very moist climatic conditions, the liverwort cover on the forest floor is apparent. In more arid environments, liverworts are rare on all substrata. The following genera are frequent inhabitants of logs in early stages of decomposition, when bark is still intact or when a log is without bark, and the liverwort establishes itself directly on the solid wood surface: *Bazzania, Blepharostoma, Calypogeia, Cephalozia, Cephaloziella, Diplophyllum, Geocalyx, Herbertus, Jamesoniella, Jungermannia, Lepidozia, Lophocolea, Lophozia, Odontoschisma, Plagiochila, Ptilidium, Scapania,* and *Tritomaria.* These genera, as can be seen by comparison, are the same as those that occupy the living tree. In many, but not all, cases, different species are involved.

As the log decomposes, it tends to form a more porous and friable substratum that makes it a moister medium for bryophyte growth. The following genera then become colonists and persist until the log is incorporated into the soil: *Anastrophyllum, Apotreubia, Barbilophozia, Bazzania, Blepharostoma, Calypogeia, Cephalozia, Diplophyllum, Geocalyx, Jamesoniella, Jungermannia, Kurzia, Lepidozia, Lophocolea, Lophozia, Mylia, Odontoschisma, Plagiochila, Riccardia, Scapania,* and *Tritomaria.* Again, those genera in common with logs in earlier stages of decomposition tend to have different species represented on the well-decomposed wood.

Occasionally thallose liverworts occupy the rotten logs, but it is not of frequent occurrence: *Conocephalum, Pellia, Riccardia,* and, occasionally, *Aneura.*

Tundra Habitats

In alpine and subalpine sites and at northern latitudes of Alaska and British Columbia, tundra habitats or herb-dominated habitats contain a number of characteristic or locally abundant liverworts or hornworts. In the tundra of the north, the mountains show the greatest diversity of liverwort species. Liverwort diversity reaches its best expression in the Brooks Range of Alaska, but also shows considerable richness in the northern Rocky and Cassiar Mountains of British Columbia. Genera of liverworts found (usually not in great abundance) only in this habitat are *Arnellia, Ascidiota, Bucegia, Calycularia, Cryptocolea, Lejeunea, Mesoptychia, Metacalypogeia, Peltolepis, Pseudolepicolea,* and *Sauteria.* Other genera widespread in Pacific North America show greatest diversity in this habitat: *Anastrophyllum, Barbilophozia, Cephalozia, Cephaloziella, Gymnomitrion, Jungermannia, Lophozia, Marsupella, Nardia, Scapania,* and *Tritomaria.*

Alpine tundra and subalpine herb-dominated sites of British Columbia and southward to California contain the same widespread genera, but a number of other genera reach greater diversity here, especially in the near coastal mountains of British Columbia and southernmost Alaska. As the suitable habitats become more restricted, as in Oregon and California, diversity is greatly reduced. Additional widespread genera that are species-rich, or are surprisingly present in these habitats, are: *Aspiromitus, Asterella, Athalamia,* and *Herbertus.* Genera restricted mainly to this zone, especially in near-coastal mountains, are *Anastrepta, Chandonanthus, Eremonotus, Haplomitrium, Schofieldia,* and *Takakia.*

Most liverworts and hornworts do not compete well with vascular plants, and thrive best in sites before the vascular plants invade. It should be noted, however, that some vascular plants create shaded sites in which bryophytes thrive. When the vascular plant vegetation is destroyed, the hepatics also disappear. The bryophytes, through alteration of the site, generate a more water-retentive substratum that is congenial to vascular plants. In consequence the hepatics are best represented on open disturbed sites, including late-snow areas, margins of watercourses, cliffs, and near boulders. Some liverworts persist at the bases of vascular plants, but they tend to develop best away from such plants.

Seasonality of Liverworts and Hornworts

Most leafy liverworts are perennial and, although some dry out during dry periods, they can be seen throughout the year. The genus *Fossombronia* is an exception, surviving the unfavorable season mainly as spores. The hornworts are decidedly seasonal in their appearance because the thallus dries out and shrinks to invisible perennial fragments or survives as spores. Some thallose liverworts are clearly annual; most species of *Riccia* survive the unfavorable season as spores. Other thallose liverworts, some species of *Asterella* and *Athalamia,* for example, survive as small fragments of apices of branches while the rest of the thallus disintegrates. Other thalli, for example, *Mannia* and *Targionia,* curl up and revive with the reappearance of favorable moisture conditions.

The genera *Sphaerocarpos* and *Riella* survive the unfavorable season as spores, and are therefore annual. *Geothallus,* on the other hand, is perennial as tubers made up of small portions of apices of thallus lobes. It also can be annual, surviving the unfavorable season as spores.

Growth of most perennial liverworts is probably slow, although branches of *Conocephalum* in favorable sites probably elongate by more than 10 cm each year. Most liverworts probably elongate less than a centimeter each year. In seasonably dry sites growth appears to be extremely slow, probably only 2–3 mm each year. Even in wet sites some liverworts grow very slowly. *Riccardia*, for example, sometimes elongates less than 5 mm each year.

Distribution Patterns of Liverworts and Hornworts in the Region

The area of Pacific North America that has been most thoroughly explored is the coastal portion, including the most accessible mountains. This is also the area that contains the greatest diversity and luxuriance of liverworts and hornworts.

At the generic level there is a certain degree of restriction of geographic range only when the genus has a single species or has very few species. A few genera (e.g., *Aneura, Blasia, Conocephalum*) that contain a single species are widespread. Most genera that contain more than a single species show a wide distribution in the area.

Genera that show a more restricted distribution can be related to specific ranges:

1. Predominantly near-coastal in the southwestern part of British Columbia extending southward west of the Cascade Mountains and Sierra Nevada to southern California: *Anthoceros, Aspiromitus, Lunularia, Riccia, Sphaerocarpos,* and *Targionia*. Two genera are restricted, in the region, to California: *Geothallus* and *Riella*; the genus *Cryptomitrium* is predominantly Californian.

2. Predominantly near-coastal in extremely high-precipitation areas of southeastern Alaska and adjacent British Columbia, reaching their greatest frequency in coastal areas exposed to the wet air masses moving in from the Pacific: *Anastrepta, Apotreubia, Chandonanthus, Cololejeunea, Douinia, Calycularia, Dendrobazzania, Kurzia, Mastigophora, Odontoschisma, Pleurozia, Sphenolobopsis,* and *Takakia*.

3. In more humid climates near the coast, then reappearing in moister climatic elevations of the interior mountain chains, but absent in the drier climatic areas between, from southeastern Alaska to northern California: *Bazzania, Blasia, Conocephalum, Geocalyx, Gyrothyra, Herbertus, Pellia,* and *Porella*.

4. In summer-dry climates that characterize southwestern British Columbia, and southward to southern California west of the interior mountain ranges: *Asterella, Fossombronia, Lunularia, Mannia, Reboulia,* and *Riccia*.

5. Mainly alpine or subalpine, possibly throughout the region, absent or less frequent southward, especially in the Sierra Nevada; in some instances, especially near the open coast, extending to sea level: *Anthelia, Athalamia, Eremonotus, Gymnomitrion, Haplomitrium, Harpanthus, Moerckia, Nardia, Pleurocladula,* and *Schofieldia*.

6. A number of genera occur mainly in mountains, generally at alpine or subalpine elevations, in Alaska and extending southward in the Rocky Mountains and the Columbia Mountains of British Columbia: *Arnellia, Bucegia, Peltolepis,* and *Sauteria*.

Most of the rest of the genera in the region show a wide distribution, with richest diversity and frequency in the coastal portion of southeastern Alaska and British Columbia where precipitation is high and temperatures are mild most of the year.

If single areas of liverwort diversity and abundance are sought, the Alexander Archipelago of Alaska and the Queen Charlotte Islands and Pitt Island of British Columbia, especially near their western coasts, show extraordinary luxuriance of liverwort cover and rich diversity of species. In the Brooks Range and northward in Alaska, there is also great hepatic diversity. During spring and early summer these plants show the greatest frequency of sporophytes or gemmae, but these periods coincide with the wet weather that can be daunting to a naturalist. From northwestern Vancouver Island northward in the many islands and islets as well as on the adjacent mainland, especially along the fjords northward to the Alexander Archipelago of Alaska and extending along the Aleutian Chain, the liverwort flora is especially rich. Open forest, canyon and cliff walls, terraced outcropping rock, and stream, streamlet, pond, and lake margins all yield a luxuriant cover and high diversity of liverworts in this climatically controlled vegetational region.

Dense forests, bogs and fens, and semiarid steppes tend to be impoverished in hepatic cover and diversity. The same is true for summer-dry climates, where the hepatics are apparent only in the winter and spring.

Critical Determination

Although this book is designed to treat genera of liverworts and hornworts, a number of species are also treated. In those genera in which there is a single local species, of course, one gets the species name as a bonus. In larger genera, however, it is usually necessary to refer to a more technical book to make species determinations. Regrettably, no technical book exists that treats all liverworts and hornworts known to occur in western North America. Many species in this area, however, are widely distributed in the Northern Hemisphere, and manuals provided for these other geographic regions can be consulted for their determination. Undoubtedly the most useful volumes are R. M. Schuster's *The Hepaticae and Anthocerotae of North America*. These treat all taxa known to occur in the eastern half of North America. Illustrations are exceptionally fine and the descriptions meticulously accurate. Comments are also extremely useful. The keys are somewhat difficult to use, especially for a beginner, and the specialized vocabulary can take some time to master. This is unquestionably one of the finest available treatises on liverworts.

A number of our local species that are absent in eastern North America are present in Europe. A most useful book that treats these species is S. M. Macvicar's *The Student's Handbook of British Hepatics*. Although badly out-of-date (it was published in 1926), this book contains useful illustrations, descriptions, and discussions of many of our species. The recent (1999) book by Jean A. Paton, *The Liverwort Flora of the British Isles*, is also an invaluable reference for our flora. More than 70 percent of our species are treated and provided with numerous excellent figures as well as a detailed treatment of the features that make the genera and species identifiable. It is an immensely scholarly work and is very "user friendly." Microscopic details are generally necessary for effective use of the book. Still other of our species are found in Japan, but the manuals available for these species are written in Japanese, thus inaccessible to most North Americans.

As mentioned earlier, three manuals have been published that treat the hepatics and hornworts of portions of the region. The best and oldest of these is M. A. Howe's *Hepaticae and Anthocerotes of California*, published in 1899 as a Memoir of the Turrey Botanical Club, vol. 7. Although the book contains few illustrations, the descriptions and discussions are exceptionally useful. Nomenclature is badly out-of-date. Less useful is L. Clark and T. C. Frye's *The Liverworts of the Northwest*, published in 1928. It contains descriptions of all treated species and genera and illustrations of many. It was issued as a Publication of Puget Sound Biological Station, vol. 6. Possibly the least useful publication is E. L. Sanborn's *Hepaticae and Anthocerotes of Western Oregon*, published in 1929 as a University of Oregon Publication, Plant Biology series, vol. 1(1). It serves, however, as a summary of the knowledge of the local hepatics at the time the treatment appeared. It contains keys, descriptions, and a limited number of illustrations, mainly habit sketches.

A more recent, but unpublished, extremely useful treatment is Judith Godfrey's Ph.D. thesis completed at the University of British Columbia in 1977, *The Hepaticae and Anthocerotae of Southwestern British Columbia*. This contains, besides introductory information, exceptionally well-constructed keys to genera and species and brief descriptions and comments concerning the distinctive features of these plants. It is based on a fine synthesis of the modern understanding of the hepatics founded on a wide background in field experience with living plants and careful study of these species in the laboratory. W. S. Hong has published treatments of a number of families, their genera and species, for western North America. P. G. Davison has produced a thorough treatment of Aleutian Chain hepatics that remains unpublished.

All of these treatments are not directed at the novice, and thus contain the specialized vocabulary of the discipline and presuppose a botanical background.

Latin Names

All organisms are assigned Latin names. Such names are used internationally, and each species has a unique name. Popular names in English are rare among liverworts and hornworts, reflecting the fact that most people overlook these plants.

A genus is used to categorize a highly distinctive organism that has a number of features unique to it. A genus is composed of one or more species. Species show the features that distinguish the genus from other genera, but each species has a number of features unique to it, thus distinguishing it from the other species of the same genus. An analogy extracted from inanimate objects would be the generic term "box," among which wooden box and cardboard box would be differing species.

Using a genus discussed in this book as an example, *Ptilidium* contains three local species. The genus is unique in a combination of features, including general form of the leaves, their arrangement, and the branching pattern of the stems. These shared characteristics are not known in other genera in the province, thus the genus can be distinguished readily from most other genera. An exception is the very closely related genus *Mastigophora*, which superficially resembles *Ptilidium* but has several features not found in *Ptilidium*. Each genus and species is separated by a constellation of features that give it a unique form. Some of these characteristics are microscopic and are not noted in this book.

A code of international rules of botanical nomenclature legislates that all genera and species, when recognized as new to science, must be given a name in Latin or latinized Greek. A description in Latin must accompany the first recognition of the genera and species and must be published in a readily available journal. If these rules are not followed, the name has no validity in the scholarly community. Since Latin is a "dead" language, and since it was the essential language of scholarship for many centuries, it has been molded into a very precise means to communicate the description of new taxa, independent of the many languages that can be utilized to discuss details of these newly described plants. Some scholars consider this procedure quaint and archaic, but its defenders note the great advantages of an international language for scholarship. If used wisely and with an ear to euphony, it results in meaningful and attractive names.

Implications of Liverworts to People

The word "liverwort" was coined to denote the idea that thallose liverworts, at least, were considered useful in curing liver ailments in people. The "doctrine of signatures" is based on the presumption that each organism that possesses curative properties for human illness has been "marked" by a Supreme Being with the signature of the illness it will cure. In the thallose liverworts, the plant was thought to resemble a liver, thus it was thought it must be useful in curing liver ailments. Unfortunately, there is no decisive evidence that this is true.

Liverworts have been used medicinally by the Chinese in particular. Both *Conocephalum* and *Marchantia* have been used to make ointments that are said to be soothing for skin irritations. The liverwort is crushed in vegetable oils to produce an ointment. Indeed, First Nations People of British Columbia used *Conocephalum* for such purposes.

Recently a number of liverworts have been noted to produce substances that show antibiotic properties. The genus *Conocephalum* as well as *Radula* and *Herbertus* have yielded substances that inhibit the growth of a number of bacteria and other microorganisms, lending validity that they may be useful medicinally.

A number of liverworts have very characteristic odors. *Conocephalum conicum*, for example, when crushed, produces a pungent odor. The same is true for *Jungermannia atrovirens* and *Geocalyx graveolens*. Others are peppery in taste (*Porella roellii*) or bitter (*Barbilophozia floerkei*). It is possible that these features may be significant deterrents to plant-eating invertebrates, but convincing evidence is scarce.

Some species of *Frullania* (e.g., *F. nisquallensis*) sometimes produce a contact dermatitis that is very irritating to a few people. This allergy is developed through frequent exposure to the liverwort species. Some who work in the logging industry have been forced to change their profession after developing this allergy, which, once contracted, is incurable.

In greenhouses the thallose liverworts *Marchantia* and *Lunularia* sometimes become pernicious weeds, covering the soil surface and preventing water penetration to the seed plants. Indeed, in tree nurseries the growth of *Marchantia* has sometimes resulted in the destruction of tree seedlings. It is difficult to destroy the liverwort without also destroying the seedlings. These liverworts thrive in the conditions provided for the seedlings and produce abundant vegetative reproductive structures that are efficiently distributed to new sites through the splashing water when the plants are irrigated.

Liverworts are sometimes important stabilizers of soil banks along trails. *Pellia*, for example, is often very important in stabilizing stream and trail banks, and a number of leafy liverworts, including *Jungermannia* and *Gyrothyra* are often important in stabilizing damp, sandy surfaces.

Some liverworts serve as very sensitive pollution indicators. In water bodies, aquatic species like *Scapania undulata* are very sensitive to pollutants. Liverworts that commonly grow on tree trunks (e.g., *Frullania, Radula, Porella*) are killed when air pollutants reach a certain level that can be tolerated by the trees on which they grow.

Some leafy liverworts are suggested to be relatively reliable indicators of high levels of copper in the rock substratum on which they grow (e.g., *Cephaloziella phyllacantha*).

Liverworts, thus, have modest applications to human needs, although their potential has never been assessed with any degree of concentration.

Key to Determine Genera of Liverworts and Hornworts in the Region

Use of the Hand Lens

A hand lens (12x or 15x) is necessary for the observation of many details of liverwort structure. Most university bookstores stock such hand lenses. For field use, the lens is attached to a stout cord and worn around the neck as a pendant.

Effective use of the lens requires holding it relatively close to the eye while the examined object is held in the other hand so that light is transmitted from the object through the lens to the eye. The object is held close enough to the lens to be in sharp focus. If the lens is held far from the eye and over the examined object, far less detail is visible.

To perceive details of cell shape in translucent organs, it is useful to place the wet specimen on glass (a microscope slide is especially useful) so that light can be transmitted through the plant and then through the lens. If the plant is spread out on the glass and a drop of water is added, the clarity is often considerably improved. Often this procedure is necessary to discern underleaves.

Use of the Key

A key is constructed using contrasting couplets, each of which leads to another couplet. The more appropriate choice of the couplets is selected that best describes the liverwort being identified. Ultimately this leads to a final choice. The illustration and discussion of that genus are then consulted to determine whether it resembles the object in hand. If not, one repeats the exercise, making sure that a wrong choice was not made. In some cases, the material in hand may be aberrant or may lack features needed to determine it accurately.

It must be emphasized that, especially in genera with many species and those that show considerable variation, a key can lead only to possibilities, and microscopic characteristics are needed to provide features that discriminate the genera. Species identification in liverworts is heavily dependent upon the use of microscopic characteristics. Regrettably, therefore, in using this key, a reliable answer cannot be attained except for some species in a given genus.

Key I

1. Plants thallose (with no suggestion of stem or leaves and usually flattened and strap-like).................... Key II
1. Plants leafy (with a cylindrical stem bearing discrete leaves in 2 or 3 rows) .... Key III (p. 28)

Key II

1. Thallus with pores on upper (dorsal) surface (best noted with hand lens, but each pore usually apparent as a dot in the center of a hexagonal area)........ 2
1. Thallus lacking dorsal pores.... 20 (p. 26)

2. Gemma cups on upper surface of thallus 3
2. Gemma cups absent 4

3. Gemma cups crescent-shaped .... *Lunularia*
3. Gemma cups circular in outline .... *Marchantia*

4. Sporangium emerging from apex of undersurface of thallus lobes .... *Targionia*
4. Sporangia embedded in thallus or on specialized receptacle arising on a stalk perpendicular to thallus.............. 5

5. Sporangia embedded within thallus, exposed when thallus decomposes 6
5. Sporangia on receptacle or sporangia absent 7

6. Thallus somewhat heart-shaped, floating on surface of quiet water.... *Ricciocarpos*
6. Thallus forming rosettes on fine-textured mineral soil .... *Riccia*

7. Thallus aquatic, either floating or submerged 8
7. Thallus not aquatic 9

8. Thallus floating on surface of quiet water, with many scales beneath ... *Ricciocarpos*
8. Thallus submerged beneath surface of quiet water, lacking scales beneath. *Riccia*

9. Thallus with very conspicuous hexagonal pattern of air chambers, with pores on upper surface; produces a turpentine-like odor when crushed .... *Conocephalum*
9. Thallus with less apparent pattern of air chambers with pores; does not produce turpentine-like odor when crushed 10

10. Sporophyte-bearing umbrellalike receptacle on elongate stalk, unlobed or deeply divided into 8–9 elongate lobes 11
10. Sporophyte-bearing receptacle made of rounded dome consisting of 4–5 stout lobes 12

11. Sporophyte-bearing receptacle deeply divided into 8–9 elongate lobes .... *Marchantia*
11. Sporophyte-bearing receptacle unlobed with ruffled margin .... *Cryptomitrium*

12. Each sporangium sheathed by a white tattered skirt of an involucre...... *Asterella*
12. Sporangia lacking conspicuous involucre........................... 13

13. Sporangium opening by a lid.......... 14
13. Sporangium opening by longitudinal lines, usually regularly 15

14. Thallus, when dry, with margins curling strongly inward, making it appear wormlike...................... *Mannia*
14. Thallus, when dry, flattened .... *Reboulia*

15. Dorsal pores circular in outline when viewed with a hand lens............... 16
15. Dorsal pores obscure or irregular in outline when viewed with a hand lens 17

16. Thallus composed of several layers of air chambers when viewed in cross section; pores lacking 4 fingerlike cells in the lower portion of its opening..... *Bucegia*
16. Thallus composed of a single layer of air chambers; pores with a 4-lobed extension from the walls of its lower part, making the inner portion appear shaped like a plus sign .... *Preissia*

17. Spores and sporangium pale brown or rusty .... *Athalamia*
17. Spores and sporangium blackish brown 18

18. Ventral scales lacking appendage at apex....................... *Sauteria*
18. Ventral scales with 1–2 appendages .... *Peltolepis*

19. Thallus a single cell thick, except for a midrib 19
19. Thallus mainly several cells in thickness, thicker in middle, thinning to margins . 20

20. Thallus margin divided into leaflike lobes................................. 21
20. Thallus margin not of leaflike lobes, but thallus often branched............... 22

21. Thallus with leaflike dorsal scales, one associated with each leaflike lobe; living thallus dotted with white cells. . . . . *Apotreubia*
21. Thallus lacking leaflike dorsal scales associated with lateral lobes; living thallus lacking white cells . . . . *Fossombronia*

22. Entire surface of thallus covered by short hairs . . . . *Apometzgeria*
22. Hairs confined to thallus margin and/or midrib or absent. 23

23. Thallus with central midrib with winglike blade borne laterally on both sides of midrib; plants never aquatic. 24
23. Thallus formed of a winglike blade on one side of a stem; plants submerged, erect, fixed on mud . . . . *Riella*

24. Blade translucent; stiff hairs on margin. . . . . *Metzgeria*
24. Blade opaque; margin without hairs . . 25

25. Thallus with several involucres in center, or thallus completely covered with involucres; each containing a spherical sporangium with no seta. 26
25. Thallus with involucres confined to near tips of lobes; sporangia borne on seta or hornlike 27

26. Thallus elongate; few involucres on central portion; with irregular leaflike lobes. . . . . *Geothallus*
26. Thallus rosette-like, usually less than 5 mm in diameter, usually completely covered by elongate or subspherical involucres . . . . *Sphaerocarpos*

27. Sporophytes emerging as pointed horns arising on the surface of the thallus. . . 28
27. Sporophytes not horn-shaped, usually with a seta, or if lacking a seta, the sporangium spherical. 29

28. Sporangium and spores brownish to yellowish when ripe. . . . . *Anthoceros*
28. Sporangium and spores black when ripe . . . . *Aspiromitus*

29. Margin of thallus or branches of thallus with leaflike flaps 30
29. Margin of thallus lacking leaflike flaps 32

30. Each cell of living thallus dotted by a single pale dot (apparent with hand lens); lateral flaps usually with an associated scale on the dorsal surface of the thallus . . . . *Apotreubia*
30. Cells lacking readily visible pale dots; lateral flaps without associated dorsal scale. 31

31. Whole thallus flat with alternating lateral lobes; frequently with gemma flasks and with internal dark spots . . . . *Blasia*
31. Lateral lobes leaflike, arising from a stemlike reclining portion; lacking gemma flasks or internal dark spots . . . . *Fossombronia*

32. Thallus a free-floating aquatic, either submerged or on the surface of the water 33
32. Thallus not aquatic; if in water, attached to substratum 34

33. Thallus somewhat heart-shaped, with numerous purplish ventral scales; floating on surface of water . . . . *Ricciocarpos*
33. Thallus narrow and elongate, forked, lacking ventral scales; submerged under water surface. *Riccia*

34. Thallus composed of numerous narrow branches (usually 2 mm or less wide); rhizoids few or absent . . . . *Riccardia*
34. Thallus unlobed or with few broad lobes 2–3 mm or more in width. 35

35. Thallus rigid and brittle, somewhat greasy to the touch when wet .... *Aneura*
35. Thallus soft and flexible, not greasy to the touch when wet.................. 36

36. Thallus margin strongly ruffled and somewhat undulate.................. 37
36. Thallus margin flat or very weakly ruffled; sporangium spherical ...... *Pellia*

37. Antheridial thallus with scattered scales on both surfaces of thallus, more abundant near incurved margins of apex; sporangia spherical .... *Calycularia*
37. Antheridial thallus with dorsal scales confined mainly to midline of only upper surface of thallus, with thallus margins ruffled; sporangia cylindric..... *Moerckia*

Key III
1. Each leaf a tapered cylinder arranged around the erect stem; leafy stems arising from much-branched, rootlike system....................... *Takakia*
1. Leaves flattened, in 2 or 3 regular rows............................ Key IV

Key IV
1. Lateral leaves lobed.................. 2
1. Lateral leaves not lobed........... Key V

2. Lateral leaves mainly 2-lobed 3
2. Lateral leaves mainly more than 2-lobed.................. Key VI (p. 30)

3. Lateral leaves with lobes of same size and form.................. Key VII (p. 31)
3. Lateral leaves with 1 large and 1 small lobe..................... Key VIII (p. 33)

Key V
1. Lateral leaves subopposite to opposite, sometimes fused at base on upper surface of stem................ *Arnellia*

1. Lateral leaves alternate, never fused at base................................ 2

2. Lateral leaves often with many cilia .... *Acrobolbus*
2. Leaves lacking marginal cilia........... 3

3. Lateral leaves often with shallow sinus or two teeth at apex..................... 4
3. Lateral leaves lacking sinus or 2 apical teeth................................. 18

4. Underleaves absent or so small as to be difficult to locate................... 5
4. Underleaves apparent with hand lens.. 11

5. Notch asymmetrically placed at lateral leaf apex..................... *Anastrepta*
5. Notch symmetrically placed 6

6. Deciduous, elongate, pear-shaped perianths present in summer to autumn................... *Gymnocolea*
6. Deciduous perianths absent or not pear-shaped 7

7. Lateral leaves decurrent on upper surface of stem....................... 8
7. Lateral leaves not decurrent on upper surface of stem....................... 10

8. [Time to seek the underleaves; they are present], underleaves forked 9
8. Underleaves simple.......... *Harpanthus*

9. Plants of wet or seepage sites .... *Chiloscyphus*
9. Plants of well-drained sites... *Lophocolea*

10. Underleaf often fused with leaf base .... *Nardia*
10. Underleaf not fused with leaf base .... *Jungermannia*

11. Underleaf simple 12
11. Underleaf notched or many-lobed at apex 15

12. Plants somewhat regularly pinnate .... *Dendrobazzania*
12. Plants never pinnate; irregularly branched........................... 13

13. Underleaf often fused with leaf base .... *Nardia*
13. Underleaf not fused with leaf base..... 14

14. Lateral leaf decurrent on upper side of shoot...................... *Harpanthus*
14. Lateral leaf not decurrent on upper side of shoot................. *Jungermannia*

15. Plants forked, usually with rootlike branches emerging from undersurface of shoot...................... *Bazzania*
15. Plants not forked; lacking rootlike branches from underside of shoot..... 16

16. Underleaf with broadly triangular lobes, conspicuous under hand lens .... *Metacalypogeia*
16. Underleaf with narrowly triangular lobes, barely visible with hand lens 17

17. Plants of well-drained sites, producing abundant sporophytes in early spring .... *Lophocolea*
17. Plants in seepage, or wet sites .... *Chiloscyphus*

18. Although seemingly without lobe, lateral leaf with a small swollen pocket-like lobe visible on undersurface 19
18. Lateral leaf lacking any hint of lobe or sinus............................... 20

19. Bilobed underleaves present ... *Lejeunea*
19. Underleaves absent......... *Cololejeunea*

20. Plants erect, leaves often irregular in outline; plants arising from rootlike or knobbed white underground system .... *Haplomitrium*
20. Plants usually reclining; if erect, not from rootlike system; leaves of regular outline............................. 21

21. Underleaves visible with hand lens.... 22
21. Underleaves absent or obscure with hand lens 30

22. Underleaves visible without hand lens 23
22. Underleaves visible only with hand lens................................ 24

23. Plants grayish to bluish green, usually on organic material or acidic substrata.................... *Calypogeia*
23. Plants yellowish green to dark green, usually growing in calcareous substrata................ *Metacalypogeia*

24. Undersides of stems with clusters of rhizoids associated with purplish elliptical patches on stem ..... *Gyrothyra*
24. Undersides of stems lacking purple patches 25

25. Underleaves often fused with lateral leaves........................... *Nardia*
25. Underleaves not fused with lateral leaves............................... 26

26. Underleaves bilobed................. 17
26. Underleaves unlobed 27

27. Lateral leaves decurrent on upper surface of stem............. *Harpanthus*
27. Lateral leaves not decurrent.......... 28

28. Lateral leaves mainly strongly incurved on margins; plants often somewhat fleshy and brittle when living .... *Odontoschisma*
28. Lateral leaves mainly flat, not incurved, plants not brittle when living 29

29. Plants with nearly circular lateral leaves, yellow green to orange, often with yellow green gemmae .... *Mylia*
29. Plants with lateral leaves circular to oblong, dark to bright green, gemmae absent .... *Jungermannia*

30. Lateral leaves often with many marginal teeth .... *Plagiochila*
30. Lateral leaves lacking marginal teeth ... 31

31. Plants white or whitish green through scarcity of chlorophyll; wormlike through closely overlapping leaves .... *Gymnomitrion*
31. Plants green or of other colors, not white; sometimes wormlike, frequently somewhat flattened 32

32. Plants somewhat wormlike, brownish, on humus in tundra .... *Cryptocolea*
32. Plants not wormlike, usually green or yellowish green to orange 33

33. Some lateral leaves with shallow apical sinus .... *Harpanthus*
33. Lateral leaves all lacking apical sinus .. 34

34. Lateral leaves with decurrent base on upper surface of stem, a single pleat above the decurrency .... *Plagiochila*
34. Lateral leaves lacking decurrency and central pleat 35

35. Perianth mouth with many cilia .... *Jamesoniella*
35. Perianth mouth lacking cilia .... *Jungermannia*

Key VI

1. Lateral leaves very deeply lobed, with sinus near leaf base or leaves with ciliate margins 2
1. Lateral leaves not deeply lobed, with sinus half or less the leaf length 7

2. Lateral leaves 3- or 4-lobed 3
2. Lateral leaves with ciliate margins or with extremely elongate divisions of the leaf margins 6

3. Lateral leaves 4-lobed, the lobes a single cell width throughout .... *Blepharostoma*
3. Lateral leaves 3- or 4-lobed, the lobes at least 2 cells wide at lobe base 4

4. Lateral leaves often a mixture of unlobed, 2-lobed or 3-lobed, cylindric structures, $1/2 – 1/3x$ width of stem .... *Takakia*
4. Lateral leaves uniformly 3- or 4-lobed, flattened, slightly more than, or less than, the diameter of the stem 5

5. Leafy shoots threadlike; rhizoids infrequent; leaves lobed nearly to base, often brownish or dark green ..... *Kurzia*
5. Leafy shoots usually 0.5 – 1.5 mm wide, including leaves; rhizoids usually present on lateral branch tips; leaves divided to $3/4$ length of leaf, usually pale to vivid green, never brownish .... *Lepidozia*

6. Lateral leaves with ciliate margins or with the lobes deeply divided into slender bands over half the leaf length .... *Ptilidium*
6. Lateral leaves lacking ciliate margins, but with elongate irregular teeth .... *Mastigophora*

7. Lateral leaves 3-lobed 8
7. Lateral leaves 4-lobed 11

8. Lateral leaf margins with coarse, long teeth 9
8. Lateral leaf margins lacking teeth 10

9. Plants irregularly branched or unbranched .... *Tetralophozia*
9. Plants pinnately branched, often very regularly .... *Mastigophora*

10. Lateral leaves with one lobe much larger than the others .... *Tritomaria*
10. Lateral leaves with lobes essentially equal in size .... *Lophozia*

11. Lateral leaves and underleaves of much the same size and form, with margins of lobes often recurved .... *Tetralophozia*
11. Lateral leaves differing from underleaves (if present) in size and form, margins generally plane 12

12. Underleaves present; lateral leaves with lowermost lobe bearing cilia .... *Barbilophozia*
12. Underleaves present or absent; lateral leaves lacking cilia .... *Lophozia*

Key VII

1. Lobes also bilobed 2
1. Lobes not further divided 3

2. Lobes filamentous, a single cell in width .... *Blepharostoma*
2. Lobes narrow, at least 2 cells in width .... *Pseudolepicolea*

3. Lobes confined to upper 1/6 of leaf (leaf shallowly bilobed) 4
3. Lobes making up at least 1/3 of length of leaf 12

4. Lateral leaves secund (i.e., curved to one side of stem) .... *Anastrophyllum*
4. Lateral leaves not secund 5

5. Plants white through lack of much chlorophyll, brittle when living, usually on rock, also on earth in tundra .... *Gymnomitrion*
5. Plants green or other color, not white when living 6

6. Plants purplish to dark brown 7
6. Plants green or yellow green 10

7. Rhizoids purplish, at least at base .... *Mesoptychia*
7. Rhizoids colorless throughout 8

8. Simple underleaves present .... *Lophozia*
8. Underleaves absent 9

9. Plants coppery brown, strongly wormlike .... *Gymnomitrion*
9. Plants purplish to nearly black, generally not wormlike .... *Marsupella*

10. Underleaves absent .... *Lophozia*
10. Underleaves present 11

11. Underleaves visible only with hand lens, deeply bilobed .... *Lophocolea* or *Chiloscyphus*
11. Underleaves visible only with hand lens, unlobed .... *Harpanthus*

12. Leafy shoots 0.5 mm wide or less 13
12. Leafy shoots more than 0.5 mm wide .. 21

13. Shoots grayish white with powdery or filmy surfaces .... *Anthelia*
13. Shoots lacking powdery surfaces 14

14. Lobe apices sharp 15
14. Lobe apices blunt 31

15. Lateral leaf margins toothed .... *Cephaloziella*
15. Lateral leaf margins lacking teeth 16

16. Rhizoids red or purplish, plants of usually damp or splashed horizontal surfaces near watercourses .... *Hygrobiella*
16. Rhizoids colorless, habitat various 17

17. Plants growing in acid-rich sites (siliceous rock, peatland, etc.) 18
17. Plants growing in acid-poor sites (limestone, sandstone, etc.) ... *Eremonotus*

18. Underleaves present, at least on gemma-bearing shoots .... *Cephaloziella*
18. Underleaves absent 19

19. Plants on decorticated logs .... *Anastrophyllum*
19. Plants on rock surfaces, in peat, etc. .. 20

20. At least some of the lateral leaves with the dorsal lobe slightly smaller than the ventral .... *Anastrophyllum*
20. Lateral leaf lobes consistently equal .... *Sphenolobopsis*

21. Plants brownish, wine-colored to nearly black 22
21. Plants obviously green 24

22. Underleaves almost the same size and shape as the lateral leaves ..... *Herbertus*
22. Underleaves absent or of different form or size than lateral leaves 23

23. Perianths inverted pear-shape, readily deciduous .... *Gymnocolea*
23. Perianths not readily deciduous or pear-shaped .... *Anastrophyllum*

24. Plants alpine or subalpine, terrestrial .. 25
24. Plants not alpine or subalpine, habitat various 29

25. Leaves usually distant, distinctly concave, with apices curving inward toward stem; with large bilobed underleaves .... *Pleurocladula*
25. Leaves distant or close, not strongly concave, underleaves small or absent 26

26. Leafy shoots usually 1 mm or less in diameter 27
26. Leafy shoots 2 mm or more in diameter 28

27. Lateral leaves with pointed lobes, underleaves absent .... *Cephalozia*
27. Lateral leaves with bluntish lobes, underleaves present or absent, unlobed and sometimes fused to lateral leaf .... *Nardia*

28. Plants fleshy, dark watery green .... *Schofieldia*
28. Plants, if fleshy, pale to yellow green, never watery green .... *Lophozia*

29. Underleaves deeply bilobed .... *Geocalyx*
29. Underleaves absent or unlobed 30

30. Plants confined to wet sites, sometimes on water surface .... *Cladopodiella*
30. Plants terrestrial, or epiphytic or on rocks, not aquatic .... *Lophozia*

31. Plants heavily pigmented purplish, dark brown to nearly black 32
31. Plants decidedly green, or with reddish pigments 33

32. Lateral leaves attached at an oblique angle on the stem .... *Gymnocolea*
32. Lateral leaves transversely attached, i.e., at right angles to stem length .... *Marsupella*

33. Plants with rootlike branches arising at right angles to stem, of very wet to aquatic sites . . . . *Cladopodiella*
33. Plants lacking rootlike branches, not of very wet sites 34

34. Plants often with simple underleaves fused to lateral leaf . . . . *Nardia*
34. Plants with underleaves absent, or if present not fused to lateral leaves . . . . *Lophozia*

Key VIII
1. Smaller lobe forming a small pocket at the lateral leaf base, not an apparent flap 2
1. Smaller lobe at least 1/3 the size of the longer lobe of the lateral leaf 5

2. Smaller lobe usually stalked and forming a helmet-shaped structure with the mouth facing the base of the stem . . . . *Frullania*
2. Smaller lobe neither stalked nor helmet-shaped 3

3. Underleaves absent. 4
3. Underleaves bilobed . . . . *Lejeunea*

4. Lateral leaves not shiny, cells papillose . . . . *Cololejeunea*
4. Lateral leaves shiny, cells not papillose . . . . *Radula*

5. Lateral leaves with ciliate margins . . . . *Ascidiota*
5. Lateral leaves either toothed or without teeth, never ciliate or divided 6

6. Underleaves present 7
6. Underleaves absent. 8

7. The smaller lobe on the underside of the stem, overlapping the larger lobe (like the thumb of a mitten). . . . . . . *Porella*
7. The smaller lobe on the upper side of the stem, never overlapping like the thumb of a mitten . . . . *Lophozia*

8. Smaller lobe forming an elongate pouch partly enclosed by incurved margins of larger lobe . . . . *Pleurozia*
8. Smaller lobe not pouch-like, but merely a flap 9

9. Larger lobe elongate, parallel-sided . . . . *Diplophyllum*
9. Larger lobe as wide as long, not parallel-sided. 10

10. Both lobes sharply tapered to pointed apex, with greyish appearance from the dull cuticle. . . . . *Douinia*
10. One or both lobes usually not sharply tapered to a pointed apex; if with cuticle present, never with pointed apex on either lobe 11

11. Smaller lobe resembling a broadened thumb of a mitten . . . . *Scapania*
11. Smaller lobe sharp, not resembling a thumb of a mitten . . . . *Lophozia*

Genera of Liverworts and Hornworts

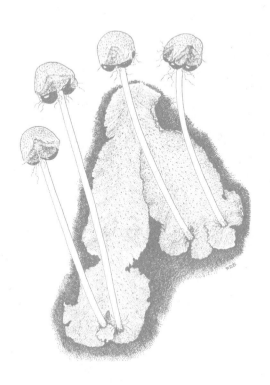

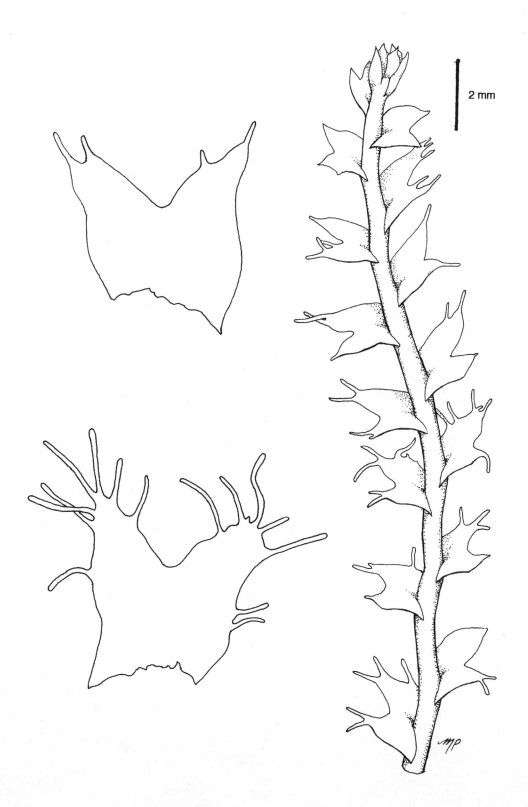

Acrobolbus ciliatus from Alaskan specimen.

Acrobolbus Nees

Name
Describes the swollen apex of the stem of the female plant where the bulblike portion is raised. Unfortunately, in the local material this structure is not produced.

Species
A single species in North America: *A. ciliatus*.

Habit
Yellowish green, flattened, creeping shoots with wide-spreading lateral leaves bearing widely spaced, distinct cilia on margins.

Habitat
Among other bryophytes on somewhat shaded rock or soil surfaces, especially near waterfalls.

Reproduction
Sporophytes unknown in the area; probably reproducing by the deciduous cilia of the leaves.

Local Distribution
Confined to Alaska: Baranof Island and Attu Island.

World Distribution
The genus is widely distributed in the tropics and subtropics, in the temperate portion of the Southern Hemisphere and extending into temperate areas of the Northern Hemisphere in East Asia and the southern Appalachian Mountains of eastern North America, and to the humid portions of the British Isles and smaller islands in the eastern Atlantic.

Distinguishing Characteristics
The yellow green flattened plants bearing shallowly bilobed leaves with distantly ciliate margins make the local species extremely distinctive.

Similar Genera
Lophocolea is superficially similar, but lacks the distant cilia on the leaf margins. Underleaves in *Acrobolbus* are rare, common (if tiny) in *Lophocolea*; *Lophozia* species with bilobed leaves are similar in form but lack ciliate margins of the lateral leaves. *Geocalyx* is also similar in habit and form, but lacks ciliate margins on the leaves.

Comments
This genus is extremely rare in western North America, as far as is known, and confined to the two localities noted.

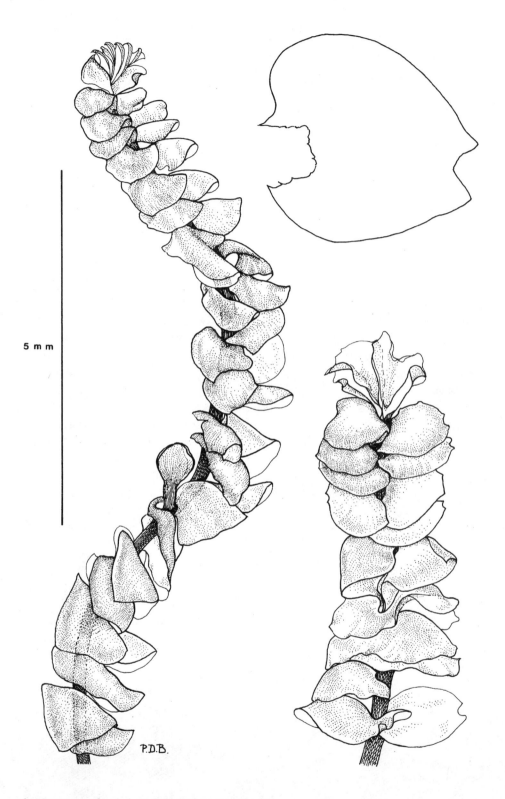

Anastrepta orcadensis from British Columbian specimen.

Anastrepta (Lindb.) Schiffn.

Name
From the Greek, meaning "twisted throughout," referring to the leaf tips facing the same side of the shoot.

Species
A single species: *A. orcadensis*.

Habit
Loose, slender, unbranched, reclining to erect shoots usually growing intermixed with other bryophytes; dull green to olive green to light reddish brown.

Habitat
Peaty banks of streams and lakes and on humus at brows of cliffs and humid terraces, often in somewhat exposed sites from near sea level to subalpine elevations, but most luxuriant in somewhat sheltered sites.

Reproduction
Sporophytes and gemmae are unknown in local material, although they are produced in other parts of its distribution in the world. Their rarity suggests that fragmentation may be the most frequent means of dissemination and may be the cause of its rarity.

Local Distribution
Confined to humid near-coastal areas of the Aleutians, southeast Alaska, and adjacent British Columbia.

World Distribution
Widely scattered, but with a very interrupted distribution in the Northern Hemisphere: in Europe from oceanic and mountainous areas of the British Isles and Scandinavia; also in the Alps, Pyrenees, and other European mountains; in Asia in the Himalayas, Japan, and Taiwan; reported from the Hawaiian Islands; in North America confined to Pacific coastal Alaska and British Columbia.

Distinguishing Characteristics
The light reddish brown to olive green shoots that usually occur sparsely among other bryophytes, in which lateral leaves are somewhat reflexed toward the ventral side of the shoot and are shallowly notched at the tip, make this a very distinctive plant. Underleaves are absent.

Similar Genera
The nearly rounded leaves with the slight notch usually separate it from any species of *Lophozia* or *Anastrophyllum*. Any *Marsupella* species of similar size tends to grow in pure, tight tufts, while *Anastrepta* tends to occur as scattered shoots or in loose turfs. *Plagiochila* may be superficially similar, but no local species of that genus has a single notch at the leaf apex (most species have many marginal teeth).

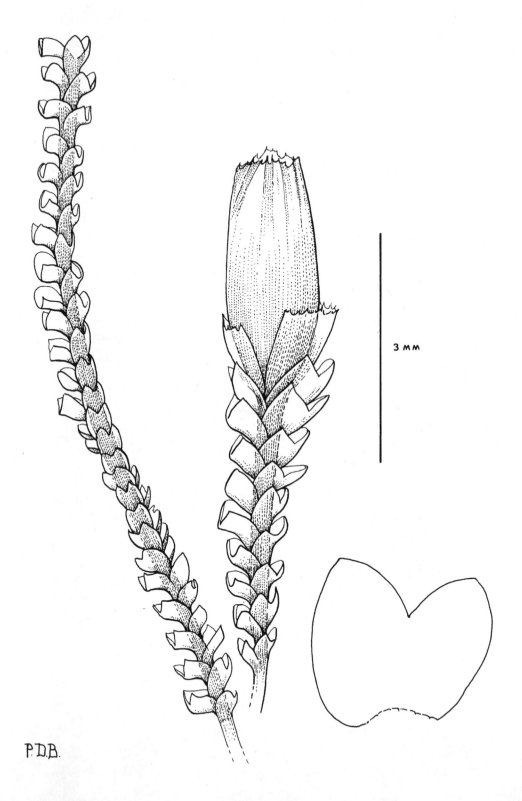

Anastrophyllum minutum from British Columbian specimen.

Anastrophyllum (Spruce) Steph.

Name
From the Greek, meaning "upward bending leaves."

Species
Eight species known in the region.

Habit
Turfs or tufts of light to dark brown to dark red-brown unbranched or irregularly branched erect plants loosely affixed to substratum.

Habitat
Varying dependent on species; on rocks (damp or dry), rotten logs, bogs, tundra, damp peaty slopes, and up tree trunks in forest, often mixed among other bryophytes; often in open sites.

Reproduction
Most species produce both sporophytes and gemmae; sporophytes tend to be most frequent in summer.

Local Distribution
The genus is widespread in the region from sea level to alpine elevations from Alaska to Washington. Its greatest diversity is in Alaska.

World Distribution
Circumpolar in the Northern Hemisphere, frequent in boreal, arctic, and montane regions.

Distinguishing Characteristics
The rusty brown to nearly black color of the plants; the usually equally bilobed leaves with a V-shaped sinus, usually less than one third the leaf length; and the usually well-drained habitat are useful characters. Although often in bogs, the plants are confined to well-drained moss hummocks.

Similar Genera
From *Herbertus*, the plants can be readily distinguished, since *Herbertus* has leaves in three rows, and its leaves are deeply two-lobed. See notes under *Tetralophozia*. From *Marsupella* it differs in the orientation of the leaves, in *Anastrophyllum* often recurved to one side of the stem, but in *Marsupella* never so.

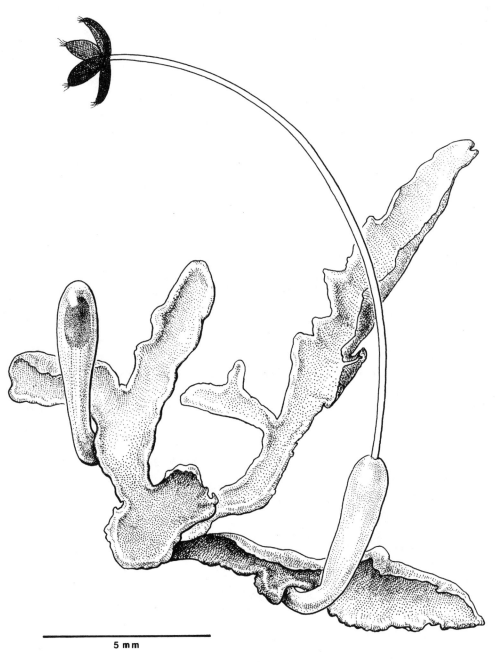

Aneura pinguis from British Columbian specimen.

Aneura Dum.

Name
Means "without a nerve," denoting that the thallus lacks a nervelike, thickened central line.

Species
One species in the region: *Aneura pinguis*.

Habit
Reclining, brittle, irregularly branched, light to dark green, somewhat greasy-textured thalli that, when crowded, become erect; often intermixed with other bryophytes.

Habitat
Damp rock, peat, or squeezed in among other bryophytes, usually in moist areas, from sea level to alpine.

Reproduction
Sporangia elongate and black, occasional, maturing in spring and summer, emerging on a white seta from an elongate, pale sleeve; the plants are brittle, thus fragments may be important in propagation. Sporophytes appear mainly in summer.

Local Distribution
Probably widespread from Alaska to California but not richly represented in collections, perhaps because plants without sporophytes are seen most frequently and not collected.

World Distribution
The genus is cosmopolitan, but reaches its greatest diversity in high-precipitation, temperate to subtropical climates.

Distinguishing Characteristics
The brittle, somewhat greasy textured (hence the name of the local species, *pinguis*), irregularly branched thalli that frequent damp sites are usually enough to separate this genus.

Similar Genera
All local species of *Riccardia* have very narrow thallus lobes and are not greasy to the touch; some specimens of *Pellia* and *Calycularia* are of similar size, but the thallus margins are notably thinner in texture than the central part, while in *Aneura*, the thallus is very thick to the margins; in *Anthoceros, Aspiromitus,* and *Blasia* the thalli show dark spots of internal *Nostoc* colonies that are absent in *Aneura*; in *Moerckia* the thallus can be similar, but it lacks the greasy texture, and thallus margins are usually thinner than the thickened central portion; in *Moerckia*, too, there are usually small scales on the upper surface of the thallus that are lacking in *Aneura*; *Calycularia* also has tiny scales, but they are on both the upper and lower surfaces, and *Calycularia* has an apparent nerve in the thallus.

Comments
The opened sporangium is unusual, with the tufts of persistent elaters at the tips of each of the divisions.

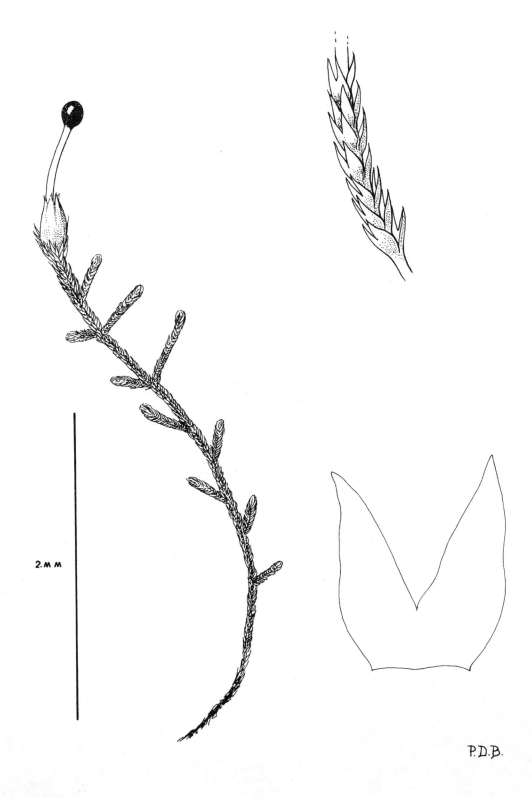

Anthelia julacea from British Columbian specimen.

Anthelia (Dum.) Dum.

Name
Means "small flower," in reference to the perianth.

Species
Two species in the region.

Habit
Forming short, hard turfs or crusts of dull grayish to silvery green, densely interwoven, slender shoots loosely attached to substratum.

Habitat
Damp peaty banks, cliffs and rocks, in shallow depressions where submerged early in the growing season, on soil of late snow areas.

Reproduction
Sporophytes occasional to abundant, maturing in summer; the seta often very short, with the spherical, dark brown sporangium sometimes barely emerging from the perianth.

Local Distribution
From near sea level on the open coast in the northern portion of its range and in alpine areas of mountains from Alaska to California.

World Distribution
Circumpolar in the Northern Hemisphere in arctic, boreal, and mountainous areas; also scattered in the Southern Hemisphere in mountainous and cool temperate areas.

Distinguishing Characteristics
The very slender, grayish plants, often coated with powdery material, the deeply divided leaves, and the wet, often alpine or subalpine habitats provide features that are usually sufficient to separate this genus.

Similar Genera
See notes under *Hygrobiella*. *Pleurocladula* also grows in alpine sites, but plants of this genus are pale green rather than powdery white, and the leaves are strongly concave with shallow sinuses. *Gymnomitrion* is often bone-white, but plants grow on well-drained rock surfaces rather than in wet sites.

Anthoceros carolinianus from Californian specimen.

Anthoceros L.

Name
Means "flower-horn," in reference to the horn-shaped sporangium.

Species
At least five species in the region.

Habit
Very compressed, dark green thalli with relatively smooth surfaces and frequent lobes; sometimes forming regular rosettes; the thalli are usually dotted with internal dark spots, visible through the surface; these are colonies of the nitrogen-fixing cyanobacterium, *Nostoc*.

Habitat
Humid shaded to somewhat open banks; on mineral soil, near sea level; in ditches and on outcrop terraces, sometimes in seasonally irrigated sites, usually apparent only in spring.

Reproduction
Sporophytes usually abundant in spring.

Local Distribution
From the southwestern portion of British Columbia southward to southernmost California.

World Distribution
Widespread throughout the world except in polar and cooler temperate climates.

Distinguishing Characteristics
It is the only hornwort in the area with brown spores, thus the sporophytes, when mature, appear brown.

Similar Genera
Vegetative material might be mistaken for *Aspiromitus*; in this genus the thalli tend to have very tattered-appearing margins and a rough surface, but the brownish sporophytes in *Anthoceros* are the most reliable feature, contrasting with the black ones in *Aspiromitus*. From vegetative thalli of *Blasia*, in which *Nostoc* colonies are also present; the thalli of *Blasia* have very regular lateral, almost leaflike, lobes; such lobes are absent in *Anthoceros* (note the figure of *Blasia*).

Comments
The species of *Anthoceros* pose problems in application of names. Material with mature spores, collected in spring, will greatly enhance the solution of these problems when it has been carefully studied. Many researchers name this genus *Phaeoceros*.

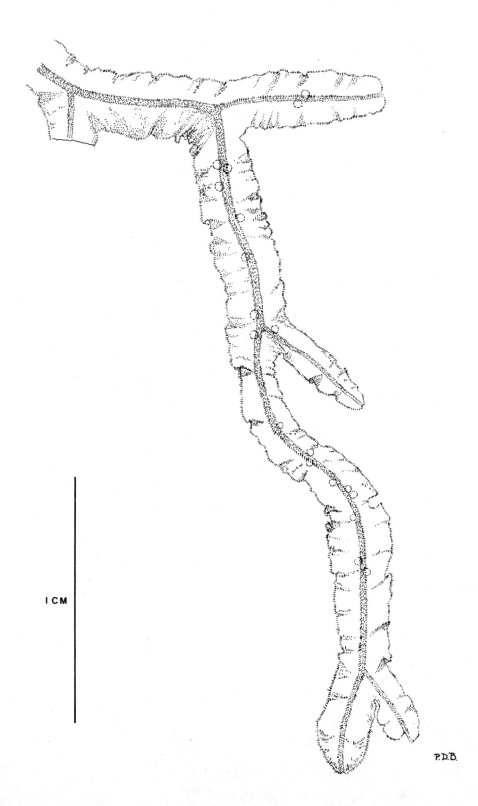

Apometzgeria pubescens from British Columbian specimen.

Apometzgeria Kuwah.

Name
Means "away from *Metzgeria*", indicating that some species were extracted from *Metzgeria* and placed into a separate genus, *Apometzgeria*.

Species
One species in the region: *A. pubescens*.

Habit
Creeping, interwoven, irregularly branched thalli, frequently on perpendicular surfaces, with apex often facing downward, forming yellow green to whitish green thin mats, sometimes mixed with other bryophytes.

Habitat
Shaded cliffs and tree trunks, usually in well-drained forested sites in regions of high humidity; rarely in tundra. Mainly in sites rich in calcium.

Reproduction
Sporophytes extremely rare, generally maturing in spring, and found most frequently on populations epiphytic on maples. Antheridial plants frequent, showing small, spherical, hairy branches near the midribs of the undersurface of the thallus. Probably also propagated by fragmentation of the thallus.

Local Distribution
Widely distributed from Alaska to northern Oregon, but most frequent near the coast; in interior regions mainly in the mountains.

World Distribution
Widely distributed in temperate portions of the Northern Hemisphere.

Distinguishing Characteristics
The somewhat irregularly branched pale green to yellow green narrow thalli that have midribs and abundant hairs on both surfaces make this a very distinctive genus.

Similar Genera
Only *Metzgeria* is likely to be confused with *Apometzgeria*, but in *Metzgeria*, hairs are rarely over the whole surface of the thallus, but are confined to the margins and midrib. Some researchers do not recognize *Apometzgeria* as a distinct genus because some species of *Metzgeria* intergrade with it.

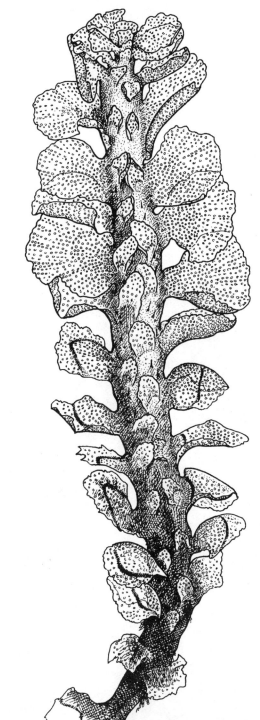

Apotreubia hortonae from British Columbian specimen.
The tiny circles represent the white oil bodies.

Apotreubia Hatt. & Kuwah.

Name
Means "away from *Treubia*," indicating that this species was removed from *Treubia* to its own independent but related genus.

Species
One species in the region: *A. hortonae*.

Habit
Creeping, brittle, pale yellow-green fleshy thalli that show, at 10x, a white dot in each cell; usually occurring as scattered plants rather than in dense colonies.

Habitat
Damp peat or humus, occasionally on rotten logs on slopes in somewhat shaded sites near sea level in extremely humid climatic areas.

Reproduction
Sporophytes extremely rare, with very short seta and spherical dark brown to black sporangia that mature in late summer. The plants are very brittle, and fragments may serve in propagation.

Local Distribution
Apparently rare, and confined to the most humid forests of near-coastal sites from northwestern Vancouver Island, British Columbia, to southeastern Alaska.

World Distribution
Southeast Asia, in mountains: the Himalayas, Taiwan, and Japan; in North America from southeastern Alaska and northwestern British Columbia.

Distinguishing Characteristics
The regular leaflike lateral lobes of the thallus, the regular placement of leaflike scales on the upper surface of the stem and adjacent to the lobes, the pale yellow-green color, and the white dots in scattered cells of the plant make this very distinctive.

Similar Genera
The very fleshy lobes distinguish this genus from *Scapania*, in which the unequally bilobed leaves are superficially similar to lateral lobes and adjacent scales of *Apotreubia*; the distinctive white dots of oil bodies are also not present in *Scapania*.

Comments
This species is the only representative of the family Treubiaceae in North America; the family is predominantly Australasian and Southeast Asian and probably originated in Gondwana, the ancient Southern Hemisphere continent that broke away from Pangaea before breaking up further to produce the Southern Hemisphere continents.

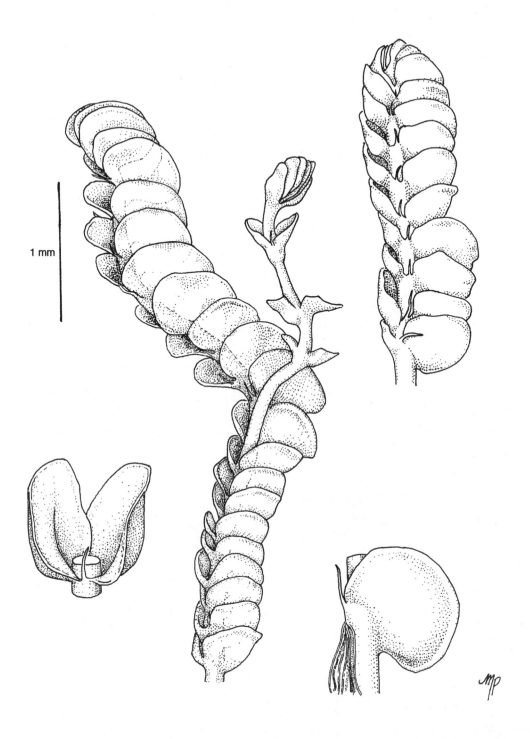

Arnellia fennica from Alaskan specimen.

Arnellia Lindb.

Name
In honor of H. W. Arnell (1848–1932), a Swedish botanist who made major contributions to the knowledge of liverworts.

Species
One local species: *A. fennica*.

Habit
Suberect light to bluish green shoots in which the leaves are nearly opposite; forming loose turfs or shoots scattered on substratum.

Habitat
On moist humus among other bryophytes of shaded to open calcareous cliff shelves.

Reproduction
Generally without sporophytes or gemmae; sporophytes mature in late summer; the perianth is toothed at the mouth and has a rhizoid-invested pouch beneath it; the sporangium is described as spherical to ovoid. Shoots are brittle, and fragments may be an important means of dispersal and reproduction.

Local Distribution
Confined to Alaska, mountainous Alberta, and the northern third of British Columbia, mainly in subalpine to alpine sites.

World Distribution
Widely distributed in the Arctic and subarctic, extending southward in mountains of temperate zones; in North America, outside the Arctic and subarctic, extending as a disjunct population in the Black Hills of South Dakota.

Distinguishing Characteristics
The pale to bluish green color, the rounded, essentially opposite leaves often fused at the base on the upper surface of the stem, and the laterally compressed shoots are very useful distinguishing characteristics.

Similar Genera
In other genera in which the plants are of similar small size, the leaves are clearly alternate and not opposite, e.g., *Cryptocolea*, *Jungermannia*, *Odontoschisma*, and *Nardia*. In *Cryptocolea*, *Nardia*, and *Odontoschisma* the leaves have a concave upper face, missing in *Arnellia*. In *Arnellia* the leaf bases can be fused on the upper surface of the stem, a feature absent in any other local genus. These genera are often difficult to distinguish using hand-lens characteristics.

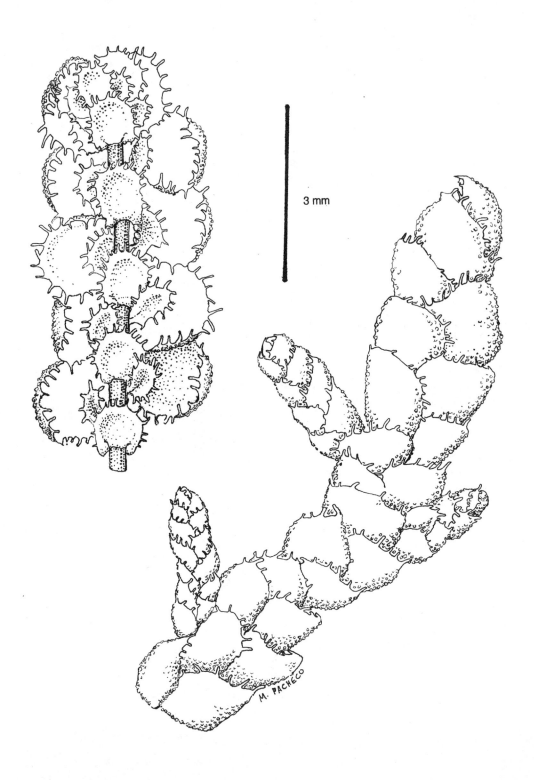

Ascidiota blepharophylla from Alaskan specimen.
The upper figure is a ventral view, while the lower shows the dorsal surface.

Ascidiota Mass.

Name
Refers to the pitcher-like perianth from which the sporophyte emerges.

Species
A single species: *A. blepharophylla*.

Habit
Forming rusty brown to nearly black prostrate patches.

Habitat
Over mosses or on bare, somewhat calcareous soil in moist depressions among sedge or grass tussocks on tundra slopes.

Reproduction
Sporophytes and gemmae not known locally, presumably disseminating by vegetative fragments.

Local Distribution
Known from a few localities in the Brooks Range of Alaska.

World Distribution
Known only from central China and from Alaska.

Distinguishing Characteristics
The very dark flattened plants with unequally bilobed leaves that bear ciliate margins combined with the tundra habitat should separate this rare genus.

Similar Genera
Porella is similar in form, but only one local species possesses teeth on the leaf margins (*vernicosa* var. *fauriei*), but this species is green, not brown to nearly black. *Frullania* is also dark in color, but the helmet-shaped lobules and bilobed underleaves as well as lack of ciliate leaf margins separate it from *Ascidiota*. *Ptilidium ciliare* occupies similar habitats, and the leaf margins are strongly ciliate, but the leaves have more than two lobes, and plants tend to be suberect rather than flattened and reclining, as in *Ascidiota*.

Comments
This distinctive genus appears to be extremely rare; most collections were made by the eminent American botanist W. C. Steere, a collector of unusual discrimination who made major contributions to the understanding of arctic bryophytes.

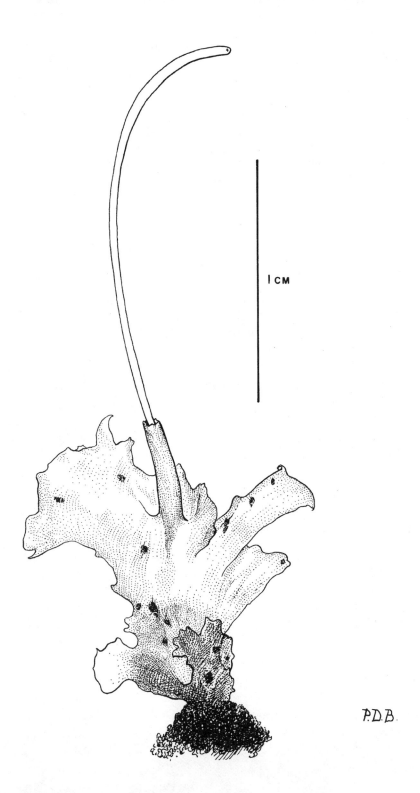

Aspiromitus punctatus from Oregon specimen.
The thallus margins are usually more irregular than shown here.

Aspiromitus Steph.

Name
Means "gentle breath," perhaps in reference to spore dispersal.

Species
Two species in the region.

Habit
Dark green flattened and ruffle-margined thalli that form rosettes firmly affixed to the substratum.

Habitat
Near sea level on disturbed earth banks of somewhat open sites, reappearing at subalpine elevations in damp ditches or on banks.

Reproduction
Sporophytes abundant in spring at low elevations and in summer at higher elevations; appearing, when ripe, as blackened, slender, erect, contorted cylinders emerging from the dark green thallus.

Local Distribution
Confined to near-coastal sites in the southwestern part of British Columbia and extending southward to southern California, where it is sometimes abundant; reappears at subalpine elevations in coastal and southeastern interior mountains of British Columbia into Washington and Oregon and southward. Also on Attu Island in Alaska.

World Distribution
The genus is widely distributed in temperate and tropical areas throughout the world.

Distinguishing Characteristics
The irregularly lobed thallus, almost circular in outline with a rough surface and dotted with small dark spots of internal *Nostoc* colonies, and the black (when mature) horn-shaped sporophytes are very distinctive features.

Similar Genera
Blasia possesses the *Nostoc* colonies, but the elongate thallus has regular lateral lobes and often bears flask-shaped gemma-containing structures that are never in *Aspiromitus*. Sporophytes of *Blasia* have a white seta and a terminal sporangium. *Anthoceros* has green sporangia with brownish to yellowish, rather than black, spores.

Comments
This genus and *Anthoceros* represent an unusual evolutionary line of bryophytes that bears very distant relationships to other bryophytes. It shows, in many ways, closer affinity to the ferns and allies. Although not convincingly represented in the ancient fossil record, this evolutionary line shows many similarities to the most ancient of the vascular plants.
In some publications this is considered to be *Anthoceros*.

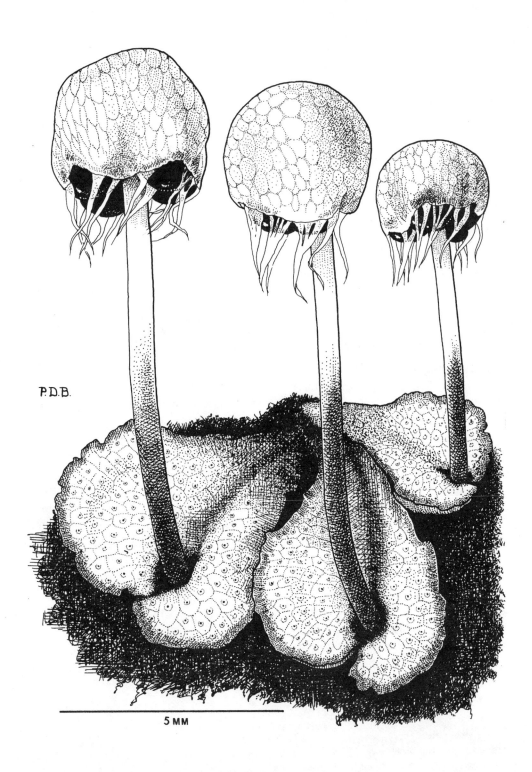

Asterella gracilis from British Columbian specimen.

Asterella P. Beauv.

Name
Means "little star," in reference to the starlike appearance of pores in the thallus of some species as viewed from above.

Species
Six species in the region.

Habit
Forming colonies, or individual plants occurring as flat rosettes of light green, rigid thalli in which the underside is dark wine-red to nearly black and the receptacles are rounded, with the four or more lobes each bearing a single sporangium sheathed by a white tattered skirt (or purplish in *A. lindenbergiana*). The receptacles can be green, as in the thallus, but are sometimes purplish brown.

Habitat
On mineral soil of open to shaded sites of terraces and ledges of rock outcrops near sea level near the coast in the southern part of its range and reappearing at alpine and subalpine elevations in similar sites as well as in forefields of glaciers.

Reproduction
Sporophytes frequent; maturing in spring at low elevations but in summer at higher elevations and latitudes. The usually white involucres around each sporophyte are striking.

Local Distribution
Found at low elevations near the coast, especially in the southwestern portion of British Columbia and extending to southern California and at high elevations throughout the region. It reaches its greatest diversity in California. *A. californica* is often common on banks and outcrops in the chaparral and oak woodland.

World Distribution
The genus is widespread from polar to tropical climates in both hemispheres, and reaches its greatest diversity in subtropical and milder temperate climates.

Distinguishing Characteristics
The rounded female receptacles with conspicuous fringes of usually white lacerate skirt surrounding each sporangium are very distinctive.

Similar Genera
All other genera that have thalli with pores and bear sporangia in receptacles lack the lacerate skirt surrounding the sporangia.

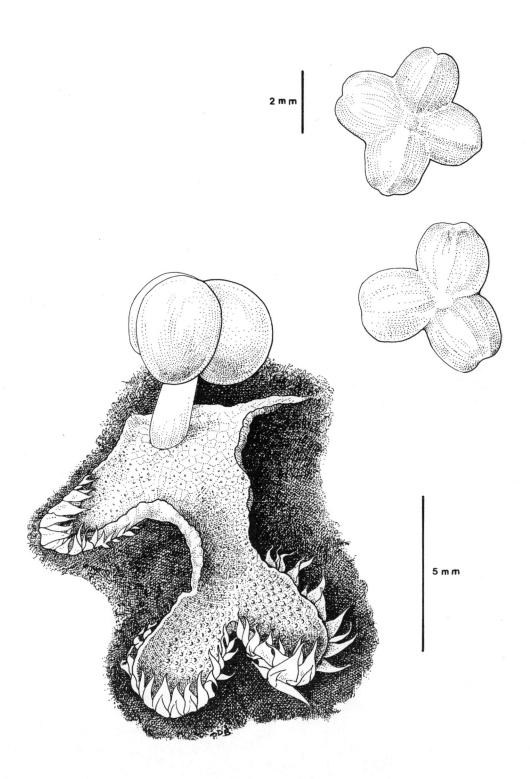

Athalamia hyalina from British Columbian specimen.
The upper right shows receptacles viewed from above.

Athalamia Falc.

Name
Means "without chambers"; although chambers are actually present, they are lost as the thallus ages.

Species
One species in the region: *A. hyalina*.

Habit
Small, brittle, light green reclining thalli with margins curved upward and fringed by white scales, especially near the tip; receptacles three- to four-lobed, each lobe bearing a single sporangium.

Habitat
Near the coast in shaded soil of cliff ledges and terraces, reappearing at alpine elevations in similar sites, especially on calcareous substrata.

Reproduction
Sporophytes frequent, with pale brown sporangia that open somewhat irregularly, maturing in spring at low elevations and in summer at high elevations and latitudes.

Local Distribution
Near sea level in the southwestern portion of British Columbia and extending southward into California; in alpine areas from Alaska southward.

World Distribution
Widely scattered in arctic regions in the Northern Hemisphere, but southward showing a broken distribution in North America, Europe, Africa, China, and South America.

Distinguishing Characteristics
The regular three- or four-lobed receptacles of the sporophytes, the irregular opening of the light brown sporangia, and the fringing white scales on the undersurface of small thalli are usually enough to separate these plants that, at low elevations, mature earlier in the spring than other similar liverworts.

Similar Genera
The features noted above separate this genus from those of similar morphology, although it is difficult to distinguish it from *Sauteria* or *Peltolepis*, especially on hand-lens features.

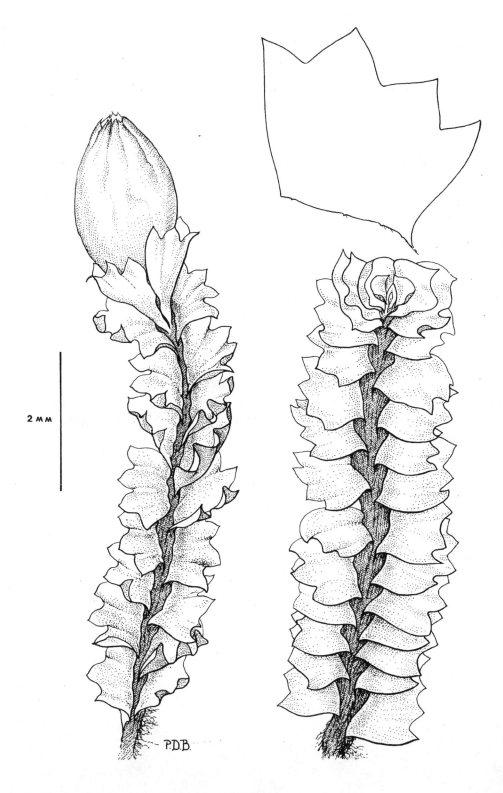

Barbilophozia barbata from British Columbian specimen.

Barbilophozia Loeske

Name
Means "bearded *Lophozia*"; the genus was segregated from the large genus *Lophozia*; in *Barbilophozia* the cilia on the leaf bases are very characteristic; the suffix *lophozia* refers to the sharp points of the leaf lobes of many species.

Species
Ten species in the region.

Habit
Plants extremely variable in size, color, and habit. Some species form turfs of erect shoots that are rusty brown to dark green while others form mats of somewhat flattened dark green shoots of reclining plants. Size varies from 2–3 mm diameter of leafy erect shoots to 1 cm diameter on leafy reclining shoots. Underleaves are usually very inconspicuous, and are best noted near shoot apices.

Habitat
Predominantly terrestrial, in subalpine to alpine to sea level; open habitats or on forest floor; a number of species are frequent on cliff ledges and at cliff bases; some extend up tree bases and are on rotten logs.

Reproduction
Gemmae are sometimes abundant, especially on margins of leaves near shoot apices; sporophytes are sometimes abundant in spring to autumn, dependent on elevation; those of low elevations appear earliest.

Local Distribution
Widely distributed throughout Alaska, southward to northern California, where infrequent.

World Distribution
A predominantly Northern Hemisphere genus, widely distributed in cool temperate climates, and extending southward in mountains.

Distinguishing Characteristics
The three to four equally lobed leaves with ciliate lower margins on one side near their attachment to the stem, the presence of ciliate underleaves, and the frequently somewhat flattened appearance of most species are characteristic features.

Similar Genera
See notes under *Tritomaria* and *Lophozia*.

Comments
Some researchers include this genus within the concept of *Lophozia*.

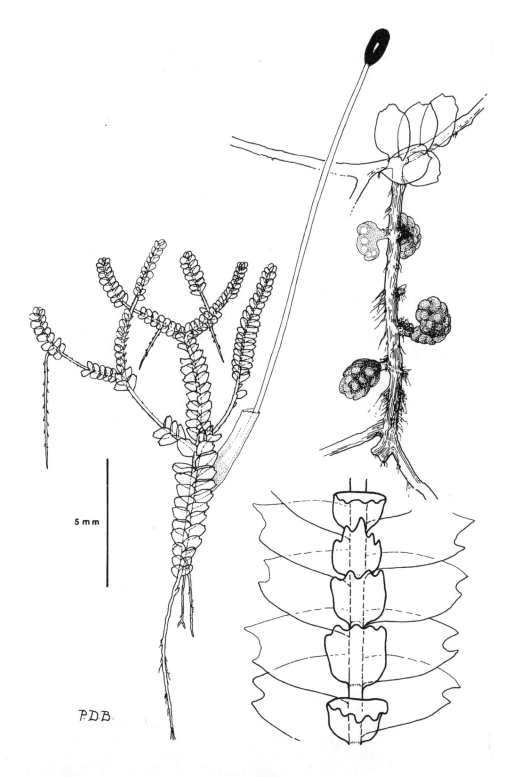

Bazzania denudata from British Columbian specimen.
Upper right is a fragment of a male plant. Lower right shows the underleaves in a ventral view.

Bazzania S. Gray

Name
In honor of M. Bazzani, an Italian anatomy professor.

Species
Four species in the region.

Habit
Loose turfs of regularly forked shoots that bear conspicuous rootlike branches that arise at right angles from the undersides of the shoots; color varying from bright yellow green, dark green, to rusty brown.

Habitat
Usually on organic substrata—tree trunks and logs within humid forest, on humus among boulders and on peaty banks, sometimes on wetland margins growing among *Sphagnum*, from sea level to subalpine elevations.

Reproduction
Sporophytes not frequent; not noted in some species in the region; in *B. denudata* maturing in spring; in this species the shoots are often denuded of brittle leaves that probably serve in propagation.

Local Distribution
Predominantly near the coast, especially in humid coniferous forests, but extending to subalpine sites, especially in Alaska and British Columbia, and less frequently southward to California.

World Distribution
The genus is essentially cosmopolitan, being absent only in arid and polar climates. In the tropics, especially in foggy highland forest, the genus is diverse and abundant.

Distinguishing Characteristics
The evenly forked branching, the often abundant rootlike branches on the undersurface of the stems, and the neatly overlapping lateral leaves that usually curve their tips downward toward the substratum added to the somewhat rounded underleaves are all very distinctive characteristics.

Similar Genera
Only *Dendrobazzania* resembles this genus, but the lateral branching in this genus is pinnate, while *Bazzania* shows forked branching. *Dendrobazzania* has rootlike branches as in *Bazzania*, but they are infrequent. *Metacalypogeia* suggests a miniature version of *Bazzania*, but this genus does not show Y-type branching and lacks the rootlike underbranches.

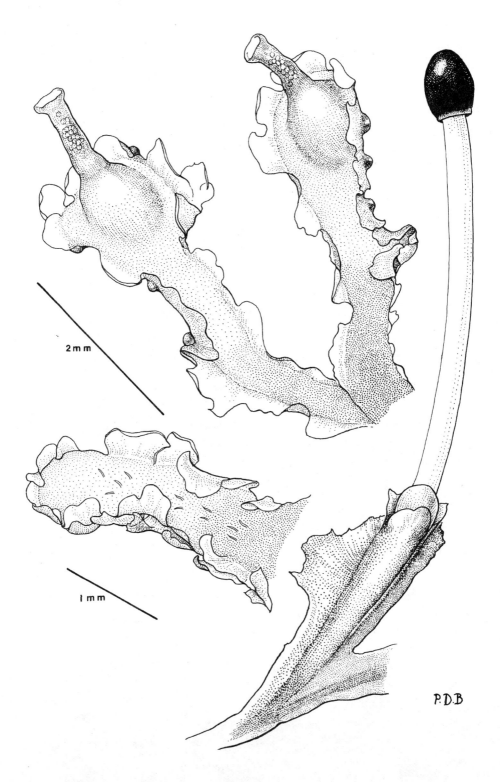

Blasia pusilla from British Columbian specimen.
Upper figure shows thallus-bearing gemma flasks.

Blasia L.

Name
In honor of Blasius Biagi, an eighteenth-century Benedictine monk of Florence and an enthusiastic botanist.

Species
One species in the region: *B. pusilla*.

Habit
Dark green thalli compressed to the substratum, the thallus usually with scattered dark internal spots visible from the upper surface, with regularly lobed and ruffled margins; plants sometimes pinkish when old.

Habitat
Disturbed humid mineral soil of roadside banks, also on sand and cliffs, especially near watercourses, in both open and shaded sites. It is occasionally abundant on recently abandoned roads in woodland.

Reproduction
Sporophytes not common, usually maturing in spring to summer; gemmae "bottles" frequent, usually apparent in spring until autumn; angled gemmae are often scattered on the upper surfaces of thalli.

Local Distribution
Widely distributed from southern Alaska southward to California, sea level to subalpine elevations.

World Distribution
Widely distributed in milder temperate portions of the Northern Hemisphere.

Distinguishing Characteristics
The very regular lateral lobes of the thallus and the frequent presence of flasklike gemma containers usually identify this genus. The dark *Nostoc* colonies that dot the thallus are also characteristic.

Similar Genera
Anthoceros and *Aspiromitus* thalli also have the *Nostoc* colonies, but these have the thalli-forming rosettes with lobes terminating the thalli rather than possessing regular lateral lobes on elongate thalli, as in *Blasia*. Sometimes *Blasia* forms rosette-like patterns, but the lobes retain their regular lobing. The presence of *Nostoc* colonies and gemma bottles separate *Blasia* from *Riccardia* or *Moerckia*. *Fossombronia* has leaflike lateral lobes from a stemlike axis, while in *Blasia* the whole thallus is clearly flattened, with no hint of a stemlike organ.

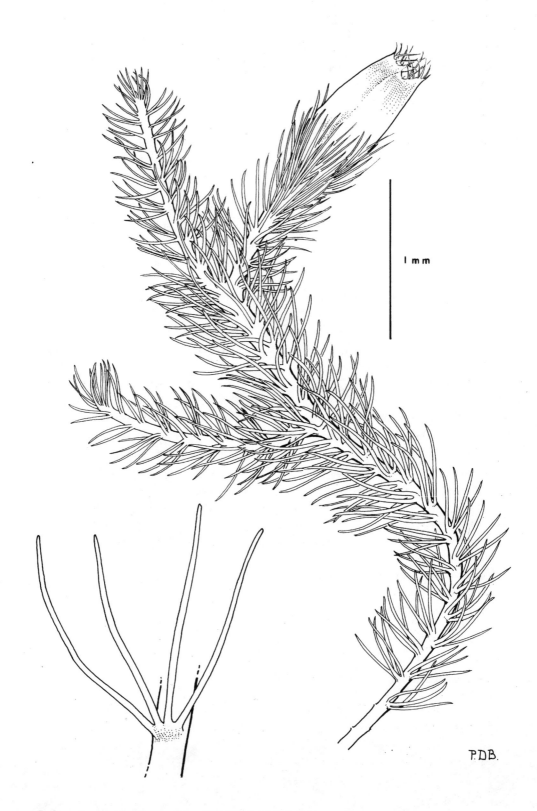

Blepharostoma trichophyllum from British Columbian specimen.

Blepharostoma (Dum.) Dum.

Name
Means "ciliated mouth," describing the mouth of the perianth.

Species
Two species in the region.

Habit
Very slender, irregularly branched shoots, yellow or pale green to vivid green, as scattered strands or as dense, short, yellow-green turfs.

Habitat
On bark of living trees, logs, among other bryophytes, in fens, and on soil, particularly on calcareous substrata, both in open sites and in forests.

Reproduction
Sporophytes relatively frequent, especially in summer; gemmae also produced near apices of shoots, replacing leaves.

Local Distribution
Widely distributed throughout much of the region from sea level to alpine elevations, but easily overlooked.

World Distribution
Widely distributed in arctic and cool temperate areas of the Northern Hemisphere, but also occasional in mountains of the Southern Hemisphere.

Distinguishing Characteristics
The tiny plants, pale to dark green to yellow green in color, the regularly four-lobed leaves in which the lobes are bristle-like, plus the ciliate perianth mouth are all useful characters. *B. arachnoideum* differs from *B. trichophyllum* in the regularly forked lobes of the leaves, while *B. trichophyllum* has unforked lobes.

Similar Genera
Takakia is sometimes similar in size, but the leaves are elongate cones rather than composed of filaments and four-lobed, as in *Blepharostoma*. *Kurzia* is similar in size, but it is brown to dark green and regularly branched, while *Blepharostoma* is usually light green and irregularly branched. *Lepidozia* has deeply lobed leaves, but the lobes are from a broad base, making each leaf resemble a tiny hand, while the leaves of *Blepharostoma* have lobes made of filaments a single cell in width. *Pseudolepicolea* is brownish and the lobes, although forked as in *B. arachnoideum*, have the lobes at least two cells wide, rather than a single cell wide as in *Blepharostoma*.

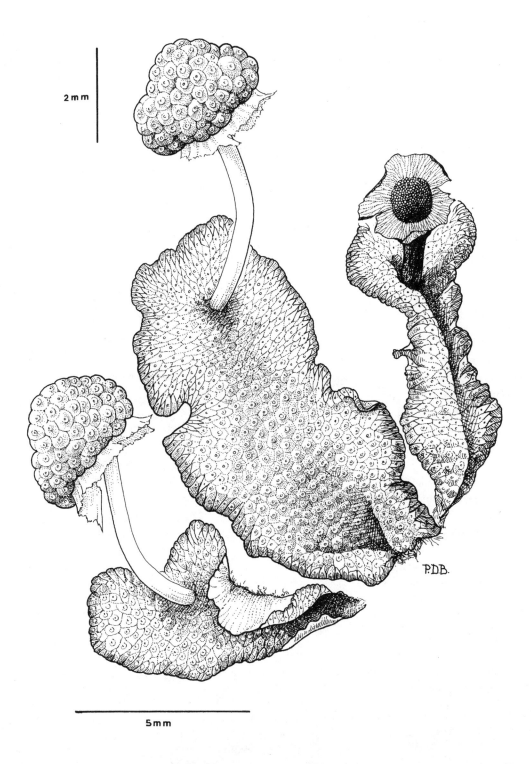

Bucegia romanica from British Columbian and Romanian specimens. The far right plant is male; the others are with female receptacles.

Bucegia Radian

Name
To commemorate the geographic area where the genus was found for the first time: Bucegi Mountains in Romania.

Species
One species in the world: *B. romanica*.

Habit
Pale green thalli firmly affixed to substratum.

Habitat
On soil of cliff shelves in subalpine and alpine areas, usually on calcium-rich substrata.

Reproduction
Sporophytes noted in the few populations discovered; maturing in summer.

Local Distribution
Extremely rare; from one locality in the Cassiar Mountains, one in Alaska, and two others in the southern Rocky Mountains of British Columbia and Alberta.

World Distribution
Known only from the Carpathian Mountains of Europe and from western North America.

Distinguishing Characteristics
The absence of four swollen cells at the base of the complex pore distinguish this genus from *Preissia*, which it closely resembles. This can be noted by viewing the pore with a hand lens. The thallus also tends to be broadly V-shaped in cross section, rather than U-shaped, as in *Preissia*.

Similar Genera
See notes above.

Comments
It is probable that this genus is often overlooked, since it closely resembles the common *Preissia*. Vegetative thalli are likely to escape notice. The world distribution is so bizarre that further collections would be welcome.

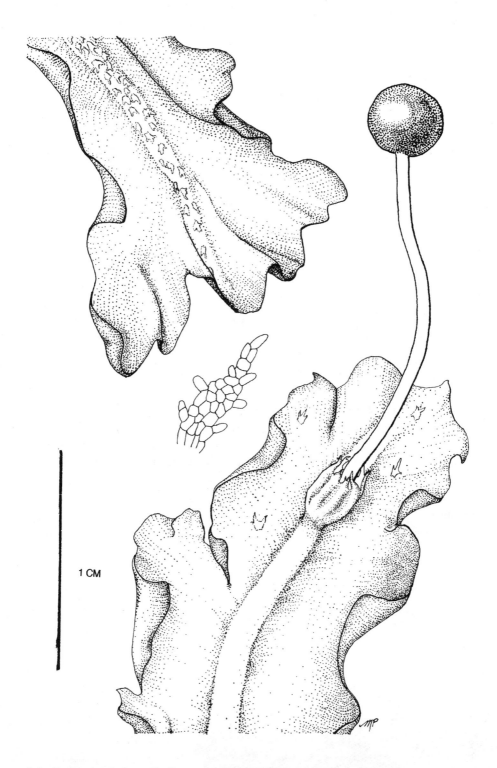

Calycularia crispula from Alaskan and British Columbian specimens and reference to an illustration by Hiroshi Inoue. The upper thallus is male. A scale is shown in detail in the center of the page.

Calycularia Mitt.

Name
Refers to the bractlike structures at the base of the seta.

Species
Two species in the region.

Habit
Light to deep green, prostrate, irregularly branched or unbranched thalli with somewhat ruffled, incurved, thin winglike margins from a thickened, riblike, lighter-colored middle portion.

Habitat
Wet acidic tundra, mixed among other bryophytes; also in shaded outcrop crevices and grottoes in alpine and subalpine regions.

Reproduction
Vegetative thalli are most often found; it is possible that fragments are disseminated; sporophytes are rare and mature in summer; the sporangium is spherical; an elongate seta may or may not be present.

Local Distribution
Known from a few localities near coastal British Columbia and in Alaska both near and distant from the coast. Since most populations lack sporophytes, the genus is neglected by collectors.

World Distribution
One Alaskan species (*C. laxa*) is known also from Siberia, while the other (*C. crispula*) is known from British Columbia, Alaska, and East Asia. The genus is represented by a third species in warm temperate to subtropical portions of Southeast Asia.

Distinguishing Characteristics
The scattered scales over much of the upper surface of the thallus, as well as similar, but fewer, scales on the undersurface of the thallus, and the lighter-colored midrib-like band combined with the somewhat incurved lobes are distinguishing features. The margin is composed of a narrow band one cell in thickness.

Similar Genera
Antheridial thalli of *Pellia* have the antheridia embedded and lack scales on the surfaces, while those of *Calycularia* are on the surface, partially protected by scales. The sporangium of *Pellia*, although spherical, as in *Calycularia*, retains a mass of elaters on the seta tip after the sporangium has opened; *Calycularia* does not. *Aneura* thalli are thickened to the margins (rather than thin-margined as in *Calycularia*), and the surface of the *Aneura* thallus lacks scales. Although *Moerckia* thalli have scales on the upper thallus surface, as in *Calycularia*, the scales are confined to the central longitudinal thickened portion of the thallus rather than scattered on both surfaces, as in *Calycularia*.

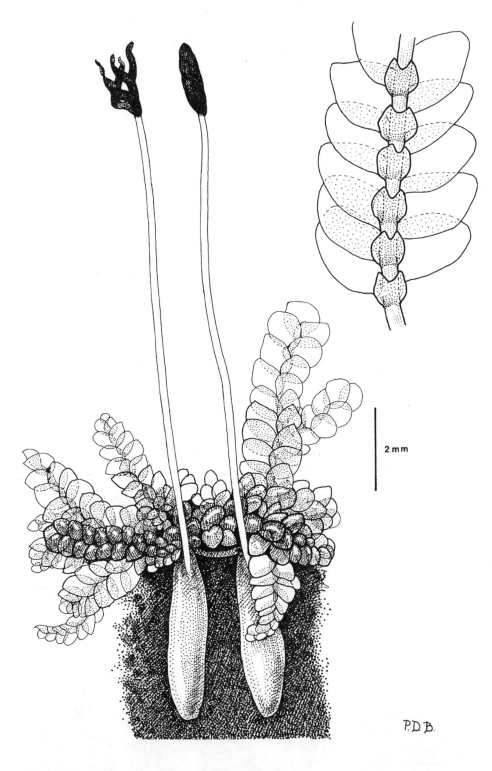

Calypogeia muelleriana from British Columbian specimen.
Upper right shows underleaves in a ventral view.

Calypogeia Raddi

Name
Means "the flower under the earth," based on the rootlike white pouch that contains the developing sporophytes.

Species
Seven species in the region.

Habit
Thin mats compressed to substratum; dark green, pale green to bluish green; usually irregularly branched, sometimes shoots suberect when in deep turfs of other bryophytes.

Habitat
On organic or mineral soil, rotten logs in deep forest, and in very wet sites, but not aquatic; sometimes growing among other bryophytes, including *Sphagnum* hummocks in peatland.

Reproduction
Sporophytes produced in spring in most species; the white marsupium (pouch) that emerges on the undersurface and burrows into the substratum is very characteristic, as are the helically arranged lines of opening of the black elongate sporangium; gemmae often abundant, especially late and early in the growing season, forming dusty pale green to yellow-green masses on upturned shoot tips. The unusual opening of the sporangium produces four long strap-shaped divisions of the sporangium wall; these uncoil rapidly when the sporangium opens.

Local Distribution
Widely distributed from Alaska southward to California, most frequent in coastal coniferous forest.

World Distribution
The genus is found in all parts of the world except in the coldest climates, from sea level to alpine elevations.

Distinguishing Characteristics
The asymmetric lateral leaves that lack lobes and neatly overlap each other like scales, the obvious underleaf, frequently with a sinus, plus the pouch from which the sporophyte emerges and the spiral lines of opening of the elongate sporangium are all useful characters.

Similar Genera
Bazzania is superficially similar, but the lateral leaves have small apical teeth; in *Bazzania*, too, the presence of rootlike ventral branches and frequent regular Y-shaped forking in the leafy branches of the plant distinguish this from *Calypogeia*. *Bazzania* has perianths rather than the pouches (marsupia) of *Calypogeia*. See also *Metacalypogeia*.

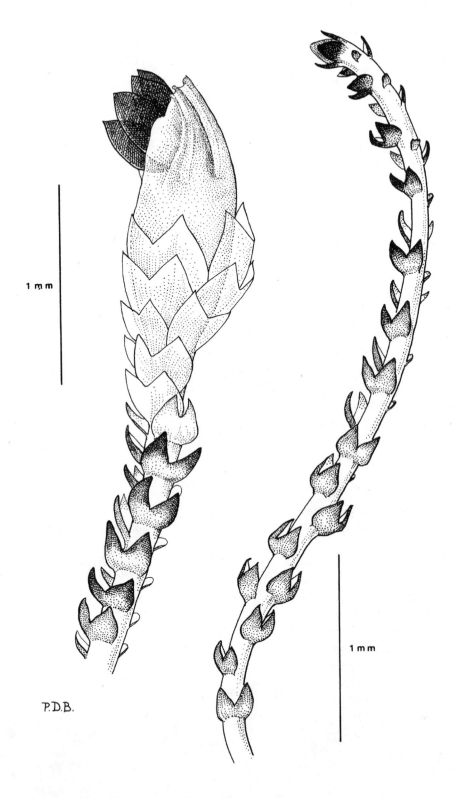

Cephalozia bicuspidata from British Columbian specimen.

Cephalozia (Dum.) Dum.

Name
Means "head-branch," referring to the usually short female branches.

Species
Nine species in the region.

Habit
Densely compressed to substratum or forming loose masses of slender, intertangled, irregularly branched, light green to dark green plants.

Habitat
Rotten logs, wet soil, peat, on living trees and among other bryophytes or low-growing vascular plants; in tundra, forests, and peatland.

Reproduction
Sporophytes common, maturing in spring and summer; those of lower elevations maturing earliest, those of higher elevations latest; usually lacking sporophytes.

Local Distribution
Widespread throughout Alaska and southward to the northern half of California, from sea level to alpine.

World Distribution
Widely distributed in the world, but mainly in temperate or cooler climates.

Distinguishing Characteristics
Most species can be placed in this genus on the basis of small size, equally bilobed sharply pointed leaves, and the consistently bright green color.

Similar Genera
Cephaloziella is most likely to be confused with this genus, but in *Cephalozia* the leaves are clearly apparent, even without a hand lens, while leaves of *Cephaloziella* are often narrower than the stem and are often difficult to distinguish, even with a hand lens. Some species of *Cephaloziella* are red brown to black, but some are green, while *Cephalozia* is green. See also notes under *Anthelia*, *Hygrobiella*, *Eremonotus*, *Pleurocladula*, and *Sphenolobopsis*.

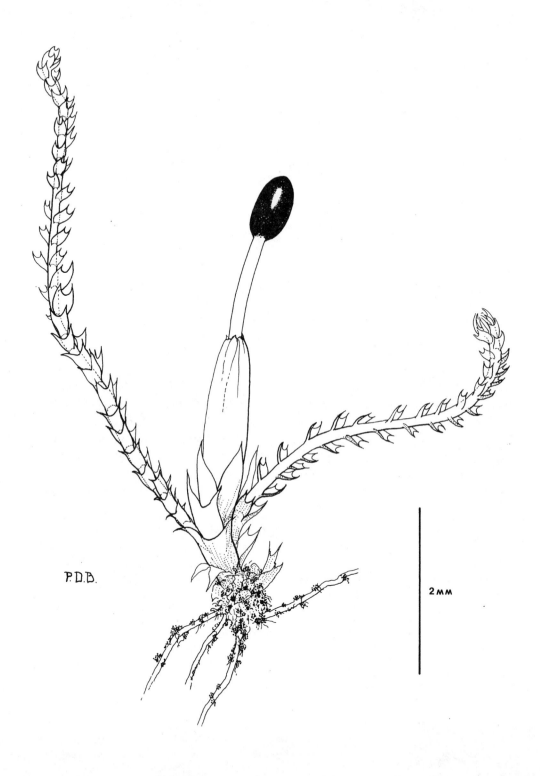

Cephaloziella divaricata from British Columbian specimen.

Cephaloziella (Spruce) Steph.

Name
Means "tiny *Cephalozia*," a satisfactory description of the plant.

Species
Seventeen species in the region.

Habit
Extremely slender threadlike strands in dense or loose masses, varying from light or dark green to red brown to nearly black; sometimes the leaves are so small that the shoots appear leafless.

Habitat
On rock or bare mineral soil; sometimes on tree trunks and dry rotten logs, often in well-illuminated sites.

Reproduction
In some species sporophytes are frequent, appearing in spring and summer. In others, sporophytes are rare; perianths are often white, especially at the mouth.

Local Distribution
The genus shows a wide distribution throughout the area in all climatic regions.

World Distribution
Mainly temperate and cool climates throughout the world.

Distinguishing Characteristics
The extremely hairlike plants with leaves barely as wide as the stem usually separate this genus, but its small size may result in considerable difficulty in recognizing patches of this plant as a liverwort.

Similar Genera
Eremonotus is similar in size and appearance, and plants are black; some species of *Cephaloziella* are also black, but those species most closely resembling *Eremonotus* grow in open areas on rock or soil, while *Eremonotus* is a genus usually of shaded cliffs or rocks. *Sphenolobopsis pearsonii* is similar in size and habitat and is nearly impossible to distinguish without a microscope. See notes under *Hygrobiella*.

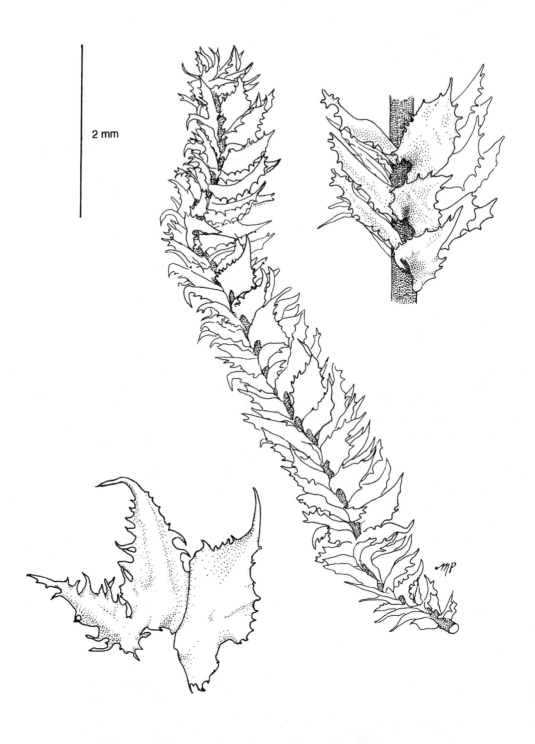

Chandonanthus hirtellus from British Columbian specimen.
The lower figure shows a single leaf.

Chandonanthus Mitt.

Name
Means "widely opened perianths"; unfortunately local material lacks perianths!

Species
One species in the region: *C. hirtellus*.

Habit
Forming loose golden green turfs of erect shoots loosely affixed to the substratum.

Habitat
Open peatland, usually in well-drained, open sites, often among other bryophytes or at the bases of shrubs.

Reproduction
Sporophytes unknown in local material; vegetative reproduction through fragmentation of brittle, dry plants.

Local Distribution
Very rare in the Queen Charlotte Islands and on Pitt Island, British Columbia, the only localities in North America.

World Distribution
Widely distributed in warm temperate and subtropical regions of Asia, also in Pacific Islands, Africa, Australia, and New Zealand; in North America only in British Columbia.

Distinguishing Characteristics
The yellowish to golden appearance of the erect shoots and the neatly three-lobed leaves with regularly toothed and recurved margins make this genus very distinctive.

Similar Genera
Only *Tetralophozia setiformis* is likely to be confused with *Chandonanthus*. In *Tetralophozia* the plants are rusty brown rather than golden. *Chandonanthus* is strictly near the coast in high humidity climates where the plants rarely dry out completely, while *Tetralophozia setiformis* is found commonly on boulder slopes where the plants withstand periods of drying. *Tetralophozia filiformis* is a species of humid rock faces, also found in humid coastal climates, but the plants are extremely slender and are rusty brown in color. Leaves of *Tetralophozia* have four lobes, while *Chandonanthus* has three.

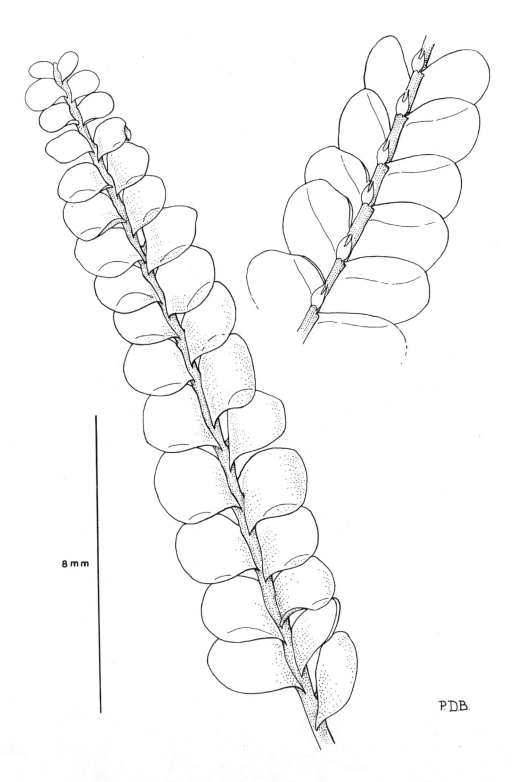

Chiloscyphus polyanthos from British Columbian specimen. Upper right shows underleaves in a ventral view.

Chiloscyphus Corda

Name
Means "lipped cup," referring to the often lipped perianth.

Species
Two species in the region.

Habit
Pallid green to dark green or yellow green irregularly branched plants, loosely affixed to substratum.

Habitat
Generally aquatic or in wet sites, often submerged or floating in masses, sometimes in running water, usually in somewhat shaded sites; sometimes on wet rock surfaces, soil, or logs and in peatland depressions.

Reproduction
Sporophytes infrequent, maturing in winter to spring at low elevations, but in summer at subalpine elevations.

Local Distribution
Widely distributed from Alaska southward to California.

World Distribution
The genus is widely distributed throughout the world, especially in warmer climates, but extends also to cooler climates.

Distinguishing Characteristics
The plants tend to be dorsiventrally flattened, the lateral leaves unlobed, underleaves strongly and deeply lobed (but very small and often obscure); color tends to be dark green, but can be whitish to yellowish green; plants are mainly in wet habitats, and are often submerged.

Similar Genera
Lophocolea is closely related, but the leaves of most species are sharp-lobed or, as in the case of *L. heterophylla*, grow in well-drained habitats. *Gyrothyra* is of similar size, but plants usually show purplish patches associated with tufts of rhizoids on the undersides of stems; in *Chiloscyphus* rhizoids are scattered and purplish patches are absent; the *Gyrothyra* usually grows on relatively well-drained mineral soil. *Jungermannia exsertifolia* grows in wet habitats, but the shoots tend to be erect and form turfs rather than mats; underleaves are absent in the *Jungermannia*. *Plagiochila porelloides* occasionally occurs in wet habitats; in this species, however, the unlobed leaves usually have toothed margins that are often recurved; this species also lacks underleaves.

Comments
Some authors consider *Chiloscyphus* to be indistinguishable from *Lophocolea*.

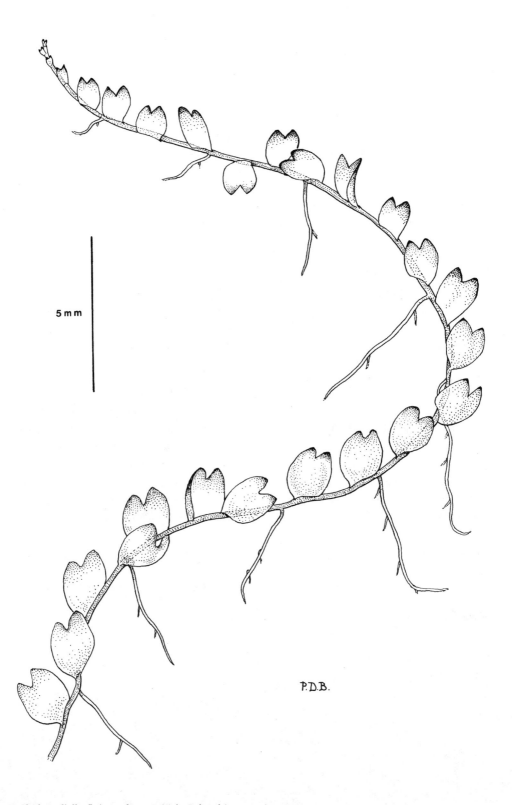

Cladopodiella fluitans from British Columbian specimen.

Cladopodiella Buch

Name
Means "tiny branch-foot" in reference to the branchlets on the rhizome-like portion of the plant.

Species
One species in the region: *C. fluitans*.

Habit
Forming dark green to purple-brown mats loosely attached to substratum or forming floating masses.

Habitat
In wet depressions and shallow pools or pool margins of peatland.

Reproduction
Sporophytes not noted in local material; plants readily fragmented.

Local Distribution
Infrequently collected, possibly rare, apparently restricted to wetlands, mainly in bogs from Alaska to Oregon.

World Distribution
Widely distributed in temperate climates of the Northern Hemisphere.

Distinguishing Characteristics
Our species is confined to boggy sites, usually aquatic or in very wet depressions; it is deep green in color, and the leaves have a shallow sinus between the lobes; flagellate branches with rhizoids and without leaves are also characteristic.

Similar Genera
Gymnocolea grows in the same sites, but plants are frequently purplish in color and lack the flagellate branches of *Cladopodiella*. In *Gymnocolea* the inflated brittle perianths are terminal, while perianths of *Cladopodiella* emerge from the underside of the shoot, but are extremely rare.

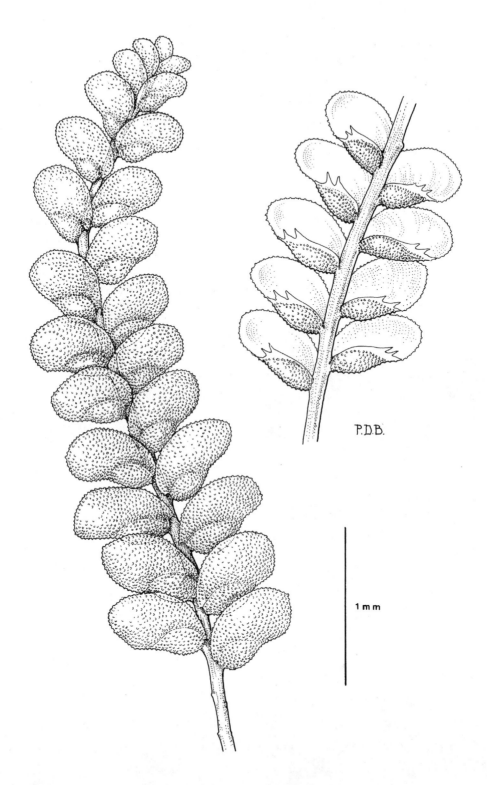

Cololejeunea macounii from British Columbian specimen amplified by reference to illustrations of R. M. Schuster. Upper right shows the ventral view.

Cololejeunea (Spruce) Schiffn.

Name
Means literally "mutilated *Lejeunea*," referring to the splitting up of the very large, mainly tropical genus *Lejeunea* into numerous segregated smaller genera, of which this is one.

Species
One species in the region: *C. macounii*.

Habit
Flattened, yellow green strands firmly affixed to the substratum.

Habitat
On bark of trees and shrubs, especially alder.

Reproduction
Sporophytes unknown in local material.

Local Distribution
Very rare; originally described from Vancouver (Hastings Mill), British Columbia, but now apparently extinct there; known also from the Queen Charlotte Islands, British Columbia, and adjacent Alaskan islands.

World Distribution
Southeast Asia, southeast Alaska, and coastal British Columbia.

Distinguishing Features
The non-glossy, yellow green flattened plants, the swollen pouchlike lobe, and the toothed lobule are distinctive features.

Similar Genera
Superficially this genus resembles *Radula* in color, size, and habitat. In *Radula*, however, the lobule is not several-toothed, leaves are glossy rather than dull, and sporophyte-bearing plants with a flattened perianth are frequent; the perianth of *Cololejeunea* is not flattened.

Comments
John Macoun collected this genus at Hastings Mill in 1889. It has not been collected since in the Vancouver area in spite of careful searching. It was noted as collected from vine maple, but all other western North American specimens are from alder; these are from southeast Alaska and adjacent British Columbia.

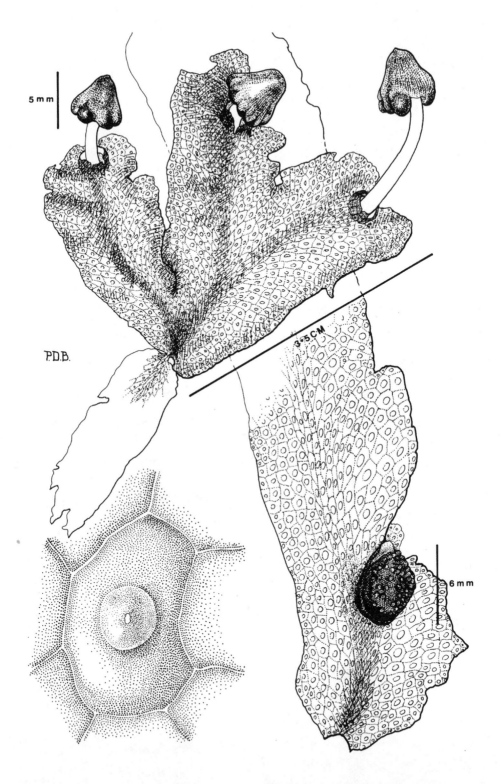

Conocephalum conicum from British Columbian specimen. Lower right shows the antheridial pad. Lower left shows the pore and one chamber outline, viewed from above.

Conocephalum Wigg.

Name
Means literally "cone head," in reference to the conical receptacles that bear the sporangia.

Species
One species in the region: *C. conicum*.

Habit
Large, light green, usually irregularly branched thalli in which the hexagonal surface outlines of air chamber pattern with central white-ringed pores are conspicuous. White rhizoids abundant on undersurface of the thallus. Sometimes plants are up to 20 cm long and 2 cm wide, especially in humid coastal climates.

Habitat
Damp depressions in open areas of woodland, on humid or wet cliffs, especially near watercourses, from sea level to subalpine elevations.

Reproduction
Male receptacles apparent in late fall and early spring, usually wine-purple pigmented. Female receptacles conic and apparent in early spring, especially when the stalk elongates very rapidly, sometimes reaching 6 to 7 cm. Sporangia black, shiny, opening by longitudinal lines to release numerous spores and elaters. Vegetative reproduction through fragmentation; internal gemmae also occur in the lowermost layers of the thallus; these are released when the layer decomposes.

Local Distribution
Widely distributed throughout the area, reaching greatest luxuriance near watercourses in moist coastal forests.

World Distribution
Widely distributed in temperate climates of the Northern Hemisphere, but infrequent in more continental areas.

Distinguishing Characteristics
The large thalli with the conspicuous outlines of air chambers, their white-rimmed pores, the aroma (like turpentine) of thalli when crushed, and the conic receptacles bearing the sporangia are very distinctive features.

Similar Genera
No other thallose genus in the area bears conic receptacles, and none shows such conspicuous air chambers or has a turpentine-like fragrance when crushed. *Marchantia* can be similar in size, but lacks the above-mentioned features of *Conocephalum*. *Lunularia* has crescent-shaped gemma cups, absent in any other local genus.

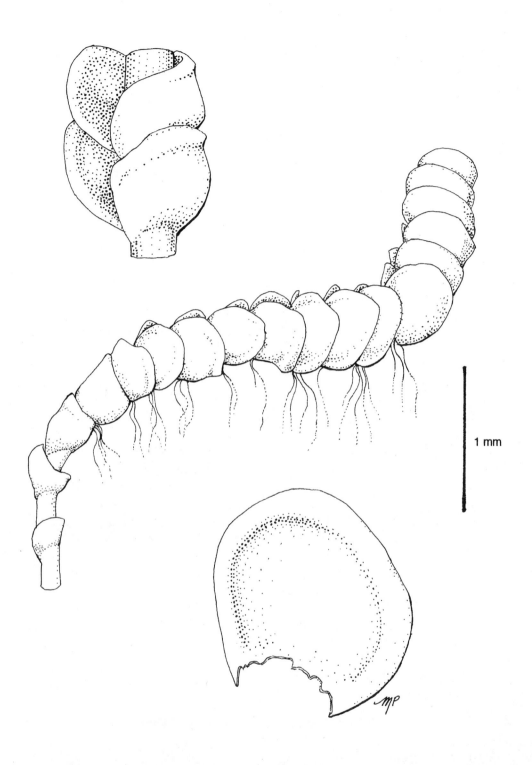

Cryptocolea imbricata from Alaskan specimen.

Cryptocolea Schust.

Name
Denotes the hidden perianth, completely enclosed by perichaetial leaves.

Species
A single species: *C. imbricata*.

Habit
Pale green to yellow green, often wormlike in appearance (from the overlapping incurved leaves), somewhat fleshy, especially the thickened creeping stems, attached to the substratum by colorless to pale brown (rarely purplish) rhizoids. Male and female plants separate, the sex organs borne terminally on the main shoots.

Habitat
Damp humus over somewhat basic rocks, usually in sheltered sites, but occurring also in open damp sites.

Reproduction
Sporangia infrequent, red brown and short ovoid when mature; perichaetial leaves often pinched to form a two-lipped structure at base of the seta.

Local Distribution
Confined to a few localities in arctic Alaska and southward to northernmost British Columbia.

World Distribution
In North America from arctic Alaska, northern British Columbia, Greenland, and Ellesmere Island in the Arctic, and with disjunct populations in the Lake Superior region of Michigan and Minnesota; also noted in the Chukotka Peninsula, Russia.

Distinguishing Characteristics
The somewhat wormlike pale green to yellowish green creeping plants with rather fleshy stems combined with the tundra habitat are usually sufficient to determine this genus. Female plants with the two-lipped mouth of the sheath of leaves at the shoot apex provide another characteristic feature.

Similar Genera
Other genera that have unlobed lateral leaves and are similar in size to *Cryptocolea* are *Odontoschisma, Jungermannia, Jamesoniella, Nardia, Arnellia,* and *Plagiochila*. *Arnellia* has opposite leaves fused at the base on the upper surface of the stem, while *Cryptocolea* and all the other noted genera have alternate leaves. *Odontoschisma* and *Nardia* have small underleaves, absent in *Cryptocolea*. *Jamesoniella*, in the region, is rarely a tundra species, and is often reddish. In *Cryptocolea* the apical leaves tend to be appressed to form a somewhat flattened structure; this is not the case in *Jungermannia* or *Plagiochila*. In *Plagiochila* with perianths, however, the resemblance to *Cryptocolea* is superficial; the perianth in *Plagiochila*, however, lacks any bracts attached to it, which are present in *Cryptocolea*. *Lejeunea* also is similar in size and gross appearance, but it has large bilobed underleaves, lacking in *Cryptocolea*.

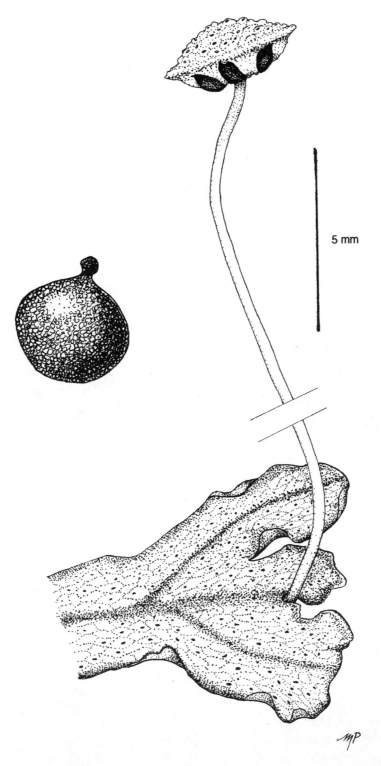

Cryptomitrium tenerum from Californian specimen aided by a photograph provided by D. Wagner. A sporangium is shown on the left.

Cryptomitrium Aust.

Name
Means "hidden turban," in reference to the inconspicuous sheath around the immature sporangium.

Species
A single species in the region: *C. tenerum*.

Habit
Pale green, flattened, dichotomously branched thin thalli, sometimes somewhat shiny, and purplish beneath, 0.6 to 1.5 cm long, less than 1 cm wide; pores small and not conspicuous; margin brownish purple in patches among green patches, somewhat undulate, curling upward when dry.

Habitat
Humid, somewhat shaded banks.

Reproduction
Sporophyte-bearing receptacles unlobed, on an elongate, somewhat grooved stalk, pale throughout or brownish purple near base; the receptacle a convex-expanded disc, thinning toward the margins. Sporangia brown, nearly spherical, with very short seta, three to seven per receptacle, each opening by a lid; maturing in early spring.

Local Distribution
From Washington State southward to southernmost California, east to the Cascades and Nevada.

World Distribution
Pacific United States and Mexico, also in South America and the Indian subcontinent.

Distinguishing Characteristics
The most distinctive characteristic is the concave unlobed receptacle that is circular in outline and bears spherical sporangia that open by a lid.

Similar Genera
When without receptacles, the thallus resembles that of several genera, including *Asterella*, *Mannia*, and *Reboulia*, but the lobed receptacles of these genera immediately separate them. The thallus is rather thin and delicate, unusual for one with pores.

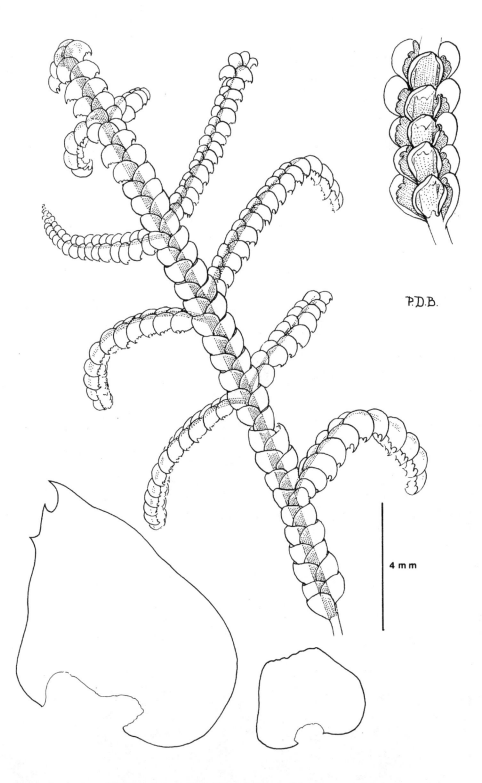

Dendrobazzania griffithiana from British Columbian specimen.
Upper right shows underleaves in a ventral view. Leaf and underleaf are illustrated at lower left.

Dendrobazzania Schust. & Schof.

Name
Means "treelike *Bazzania*," separating the pinnately branched plants from the forked-branched *Bazzania*.

Species
One species in the region: *D. griffithiana*.

Habit
Pale green to sandy brown, suberect or reclining, distantly pinnate-branched shoots, growing among other bryophytes or occurring as loose mats.

Habitat
Terrestrial in extremely humid climatic areas, usually on terraces near watercourses, from near sea level to subalpine ridges.

Reproduction
Sporophytes unknown; presumably reproducing by fragmentation of the somewhat brittle shoots and branches.

Local Distribution
Known from only a few localities in the Queen Charlotte Islands, British Columbia.

World Distribution
Known from only the Himalayan Mountains in Asia and the Queen Charlotte Islands, British Columbia.

Distinguishing Characteristics
The regular pinnate branching associated with the asymmetric lateral leaves with toothed tips and the conspicuous simple underleaves (as viewed by a hand lens) usually separate this rare plant from other local genera.

Similar Genera
Radula auriculata is similar in size, but underleaves are lacking, and the lateral leaf has a swollen-based lobule; *Porella* has an obvious lobule on the lateral leaf. *Bazzania* is never pinnately branched and usually has many naked rootlike branches arising on the undersurface of the shoots; such branches are infrequent or lacking in *Dendrobazzania*. *Frullania* is often pinnately branched, but the plants are dark red brown to nearly black and the lobules are helmet-shaped, features lacking in *Dendrobazzania*.

Comments
This very rare genus appears to be a survivor of an extremely ancient flora. Its reproduction is by unspecialized vegetative fragmentation, thus it has poor dispersal efficiency and the few available populations in the world possibly represent a portion of a flora that long preceded the evolution of humans.

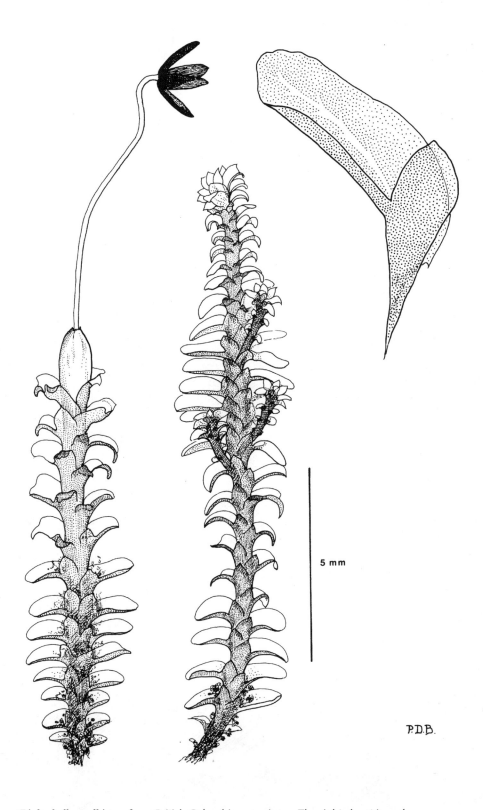

Diplophyllum albicans from British Columbian specimen. The right shoot is male.

Diplophyllum (Dum.) Dum.

Name
Means "doubleleafed," in reference to the large lobule that resembles a leaf, making each leaf appear to be a pair.

Species
Six species in the region.

Habit
Dark green to pale yellow green, sometimes purplish to pinkish tinged, glossy or dull, of suberect to reclining little-branched or unbranched somewhat flattened plants, usually loosely affixed to substratum.

Habitat
Humid cliffs, on soil, logs, rocks, or boulder slopes, in tundra, or as an epiphyte on tree trunks or bases, from sea level to alpine.

Reproduction
Sporophytes frequent in some species, but rare in others, appearing in spring or summer; gemmae frequent near shoot apices and on leaf margins of some species; fragmentation common in most species.

Local Distribution
Some species widely distributed in Alaska and British Columbia, others confined to one climatic region from Alaska to California; still others local.

World Distribution
The genus is predominantly in temperate to arctic portions of the Northern Hemisphere, with a few species in mountains of the tropics and in the Southern Hemisphere.

Distinguishing Characteristics
The usually overlapping dorsal lobule combined with the lack of underleaves is usually enough to identify the genus, but see notes cited below.

Similar Genera
See notes under *Scapania, Douinia,* and *Lophozia.*

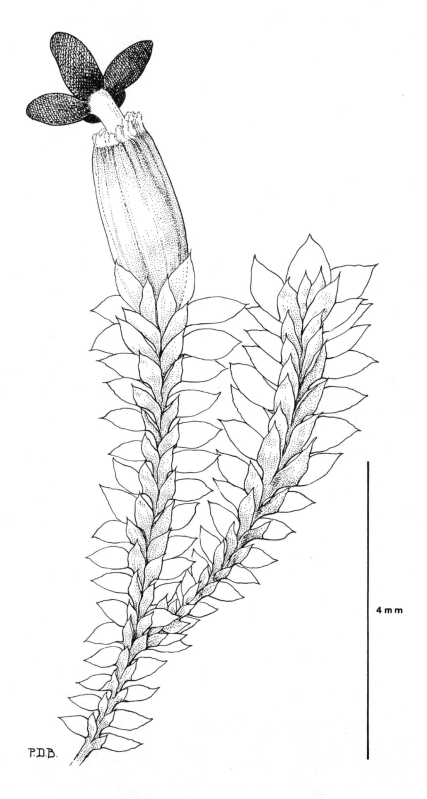

Douinia ovata from British Columbian specimen.

Douinia (C. Jens.) Buch

Name
In honor of C. I. Douin (1854–1944) a distinguished French botanist who made major contributions to the understanding of bryophytes.

Species
One species in the world: *D. ovata*.

Habit
Creeping to suberect dull bluish green to purplish brown thin mats or short turfs.

Habitat
On open to shaded cliff faces and boulder surfaces, also epiphytic on trunks of both evergreen and broadleafed trees.

Reproduction
Sporophytes frequent, especially in spring and summer; readily fragmented when dry.

Local Distribution
Confined to near-coastal regions from near sea level to subalpine elevations from southeastern Alaska to the northern portion of California.

World Distribution
Near the coast in western Europe and southern Greenland, besides its western North American distribution from southeastern Alaska southward to California.

Distinguishing Characteristics
The powdery bluish green color (often with a hint of purple), the very sharply pointed lobe and lobule, the white-tipped pleated perianth, the well-drained, usually dry habitat, and the small size are characteristic.

Similar Genera
See notes under *Scapania* and *Diplophyllum*; usually the combination of sharply pointed lobes and lobules, the small size, and the well-drained habitat are enough to separate *Douinia*.

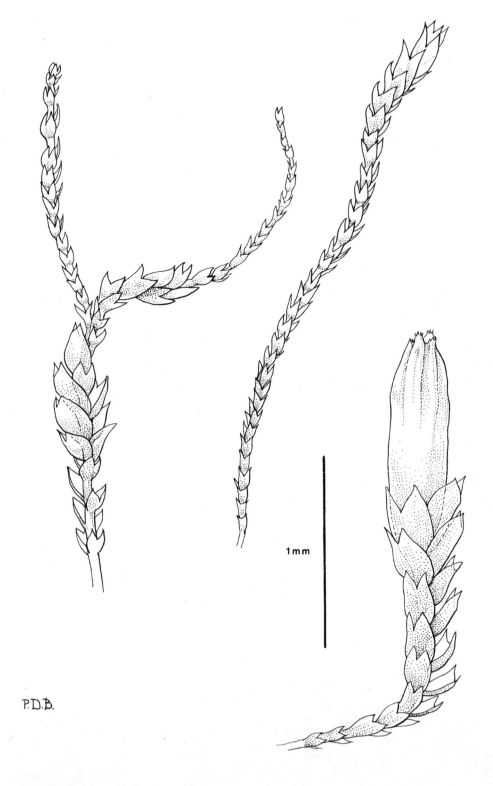

Eremonotus myriocarpus from Washington State specimen.

Eremonotus Lindb. & Kaalaas in Pears.

Name
Means literally "peaceful moisture," in reference to the humid atmosphere of the sites where the plant often grows.

Species
One species in the region: *E. myriocarpus*.

Habit
Forming thin black mats of irregularly branched shoots.

Habitat
In somewhat shaded crevices and faces of outcrops and boulders, also in flushes, generally at higher elevations, but extending to near sea level in extremely humid climates.

Reproduction
Sporophytes occasional in summer.

Local Distribution
Mainly near the coast at both high and low elevations in Alaska southward to Washington.

World Distribution
Scattered in Europe, Greenland, western North America, Japan, and China.

Distinguishing Characteristics
The reddish brown to black color of the extremely small plants with deeply bilobed leaves combined with the moist habitat are useful characters.

Similar Genera
See notes under *Cephaloziella*, *Sphenolobopsis*, *Marsupella*, and *Hygrobiella*.

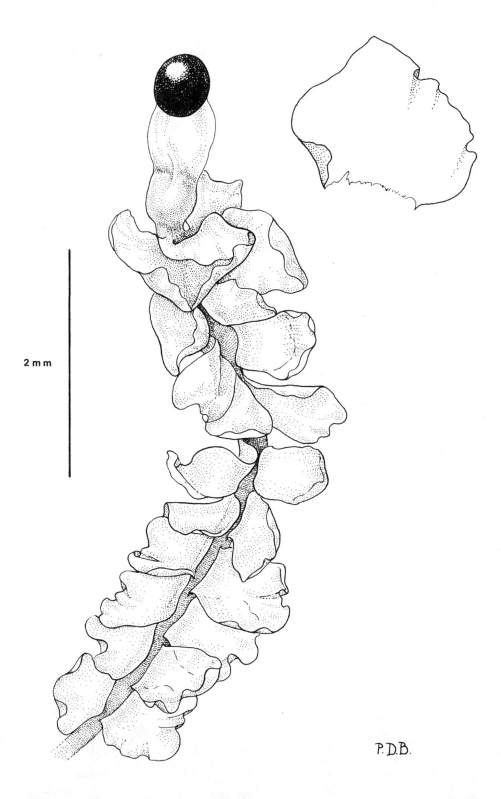

Fossombronia longiseta from Californian specimen.

Fossombronia Raddi

Name
In honor of V. Fossombroni, an Italian politician of the late eighteenth and early nineteenth centuries who supported botanical researchers.

Species
Four species in the region.

Habit
Light to dark green creeping plants firmly affixed to the horizontal substratum.

Habitat
Terrestrial on mineral soil of open areas, damp soil on lake and pond shores subject to seasonal inundation, and soil of open terraces of outcrops in the oak woodland.

Reproduction
Sporophytes frequent, in some species maturing in spring, in others in summer or autumn.

Local Distribution
Near-coastal at low elevations from northern Alaska to southern California, also in more continental portions of California.

World Distribution
Widely distributed in milder temperate climates throughout the world, but some species also in colder climates.

Distinguishing Characteristics
The most characteristic features are the leaflike ruffled lobes of the thallus that are attached longitudinally to the stem plus the spherical sporangia that open irregularly rather than by longitudinal lines. The plant is a thallus with slightly overlapping leaflike lobes. All species die down during the non-growing season in contrast to leafy liverworts.

Similar Genera
No thallose genus is similarly lobed nor are leaves of leafy genera longitudinally attached. *Lophocolea*, *Chiloscyphus*, and *Gyrothyra* that grow in similar habitats have obscure underleaves, and in these genera the leafy shoots persist throughout the year as leafy plants.

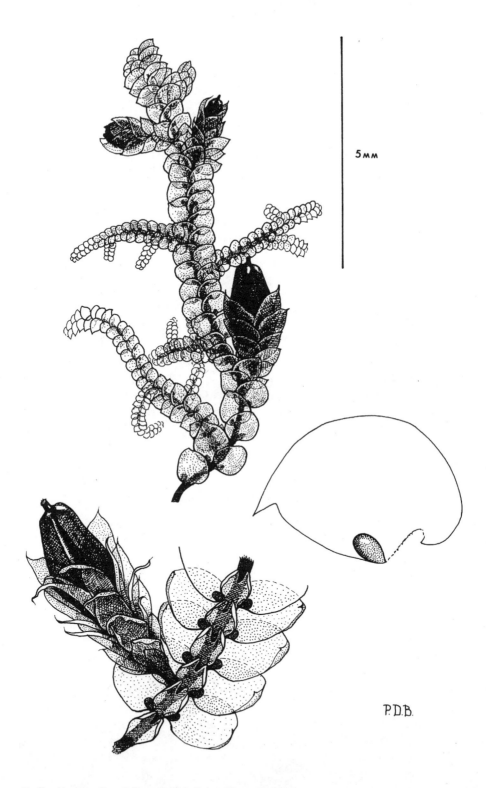

Frullania nisquallensis from British Columbian specimen. Lower left shows underleaves and perianth in ventral view.

Frullania Raddi

Name
In honor of L. Frullani, an Italian politician of the eighteenth and nineteenth centuries.

Species
Nine species in the region.

Habit
Dark glossy red brown to purplish brown, often regularly pinnate-branched plants tightly affixed to the substratum or forming suberect shoots in turfs loosely affixed to the substratum.

Habitat
Often on perpendicular surfaces as an epiphyte on trunks and branches of living trees and shrubs, on rock surfaces and on open ledges, from sea level to subalpine and alpine elevations.

Reproduction
Some species produce sporophytes abundantly in spring, and these persist through the year; others are vegetative, especially when terrestrial. The plants are often brittle when dry, and therefore easily fragmented, a probable means of reproduction, especially in terrestrial plants.

Local Distribution
Predominantly near the coast; most species west of the interior mountains from northern Alaska to southern California.

World Distribution
The genus is widely distributed in temperate and tropical areas throughout the world.

Distinguishing Characteristics
The red brown to purplish brown plants and the very distinctive helmet-shaped lobule that forms a pocket are very diagnostic.

Similar Genera
Porella plants are usually twice or more the size of *Frullania*, but the lobule is flat, not helmet-shaped; the color can be similar to *Frullania*. Some species of *Radula* may be superficially similar (e.g., *R. brunnea*), but lack underleaves and the helmet-shaped lobule. *Ascidiota*, also of similar color, has ciliate leaves and underleaves lacking in *Frullania*.

Comments
A few people, especially loggers, have become allergic to *Frullania*. It produces a severe contact skin irritation in such individuals. Epiphytic plants easily fragment, and the fragments affix, Velcro-like, to clothing or moist skin. The affliction results only after frequent contact with the plant and, once acquired, is incurable. The irritation is increased by exposure of the skin to sunlight, long after the allergy has been acquired.

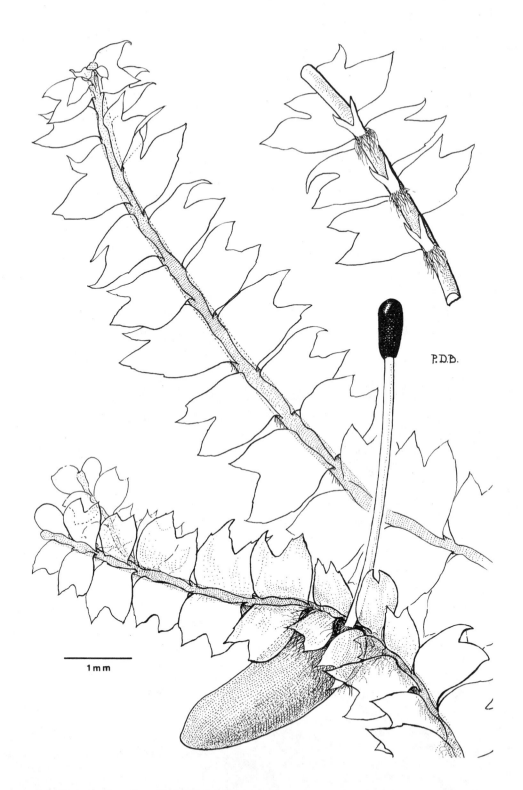

Geocalyx graveolens from British Columbian specimen.
Upper right shows underleaves in ventral view.

Geocalyx Nees

Name
Refers to the subterranean pouch that contains the developing sporophyte.

Species
One species in the region: *G. graveolens*.

Habit
Bright yellow green irregularly branched compressed plants firmly attached to the substratum.

Habitat
Usually on well-decomposed logs or on the bark of living trees in open forest, sometimes on humus.

Reproduction
Sporophytes occasional, maturing in spring.

Local Distribution
Widely distributed from southeastern Alaska southward to the northern half of California

World Distribution
The local species is widely distributed in the temperate Northern Hemisphere, while the remaining few species are widely separated in scattered localities in the Northern and Southern Hemispheres.

Distinguishing Characteristics
The bright yellow green plants, usually on rotten wood or tree trunks, the bilobed leaves, the pouchlike subterranean structure from which the sporophyte emerges, and the elongate sporangium with straight longitudinal lines of opening are distinctive features.

Similar Genera
Calypogeia has similar pouches, but lateral leaves are unlobed; *Lophocolea* can be similar in color and leaf shape, but the sporophyte emerges from a perianth and the sporangium is spherical. *Acrobolbus* is similar superficially, but the ciliate leaf margins immediately separate it. *Harpanthus* may also resemble it, but the plants are glossy, and dull in *Geocalyx*; underleaves are bilobed in *Geocalyx*, unlobed in *Harpanthus*.

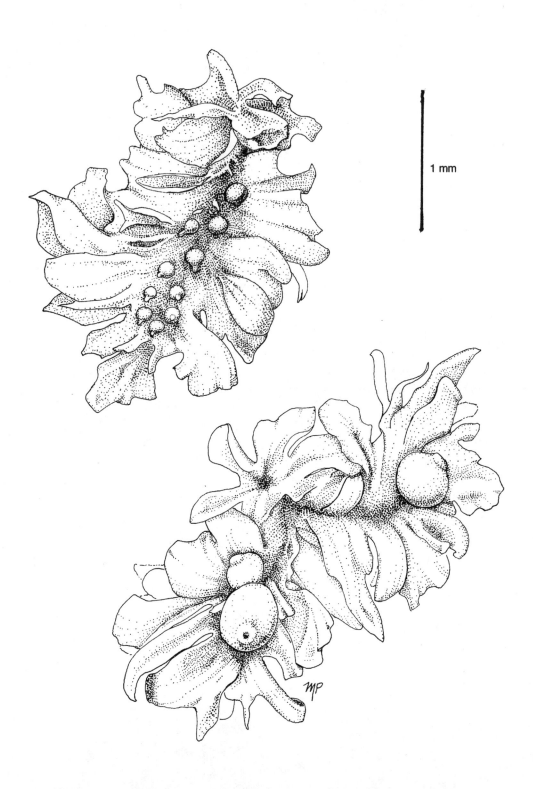

Geothallus tuberosus drawn from photographs from W. Doyle as well as from specimens provided by W. Weber. A male plant is above, a female below.

Geothallus Campb.

Name
Means "earth thallus," in reference to the plant form, firmly affixed to the soil.

Species
A single species: *G. tuberosus*.

Habit
Plants pale green, unbranched to somewhat forked, 5 to 7 mm long and 3 to 5 mm wide, with a thickened central portion and lobate leaflike wings that are flat to somewhat ascending, the thallus highly variable in form.

Habitat
Sandy soil.

Reproduction
Sporangia spherical, black, sheathed in an elongate to subspherical tube that arises on the surface of the thallus. Sporangia ruptures irregularly by decomposition or drying out of the jacket. The perennial "tubers" also survive the unfavorable seasons and germinate to produce new thalli when favorable conditions of moisture and temperature return.

Local and World Distributions
Endemic to western North America. Known only from the southernmost area of California, where extremely rare; unquestionably threatened with extinction.

Distinguishing Characteristics
The most characteristic features are the tubular sheath that surrounds a single spherical sporangium that lacks a seta, combined with the swollen "tubers" near the apices of thalli. The thalli turn black when dry.

Similar Genera
Extremely similar to *Sphaerocarpos* and occupying the same habitat; *Sphaerocarpos*, however, lacks tubers, and the thalli tend to be more regular in outline than in *Geothallus* and do not turn black when dry. *Fossombronia*, especially young plants, have a superficial resemblance to *Geothallus*, but the sporophytes, when mature, have a seta.

Comments
This genus is extremely rare and may be approaching extinction. The remaining populations are endangered by housing developments.

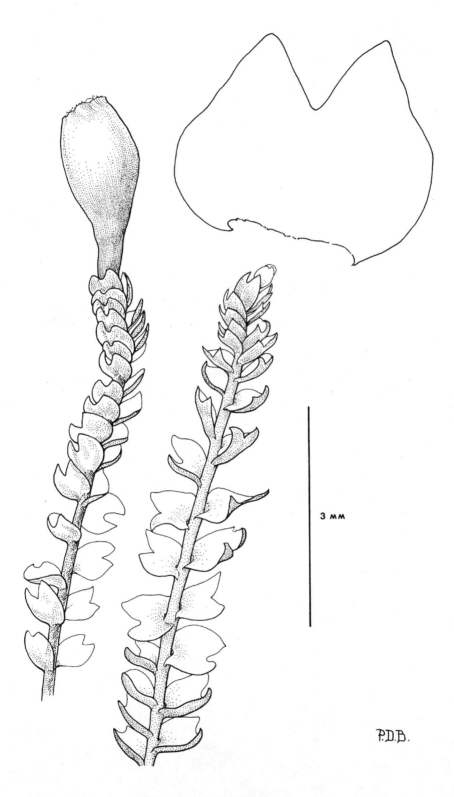

Gymnocolea inflata from British Columbian specimen.

Gymnocolea (Dum.) Dum.

Name
Means "naked sheath," in reference to the fully exposed perianth.

Species
Three species in the region.

Habit
Dark green to purplish to nearly black mats or short turflike masses; plants irregularly branched or unbranched; elliptic to pear-shaped perianths readily deciduous.

Habitat
Damp depressions at bog margins, covering basins of temporary ponds in open forest.

Reproduction
Sporophytes have not been noted in local material although perianths are common. The plants are very brittle.

Local Distribution
Widely distributed from Alaska southward to California.

World Distribution
Predominantly in arctic and cool temperate climates of the Northern Hemisphere, but two species in the Andes and another in Kerguelen Island.

Distinguishing Characteristics
The wet or boggy habitat, the rounded bilobed leaves, the usually purplish to nearly black plants, and the brittle, inverted, swollen, pear-shaped perianths are distinctive characteristics.

Similar Genera
See notes under *Cladopodiella* and *Marsupella*.

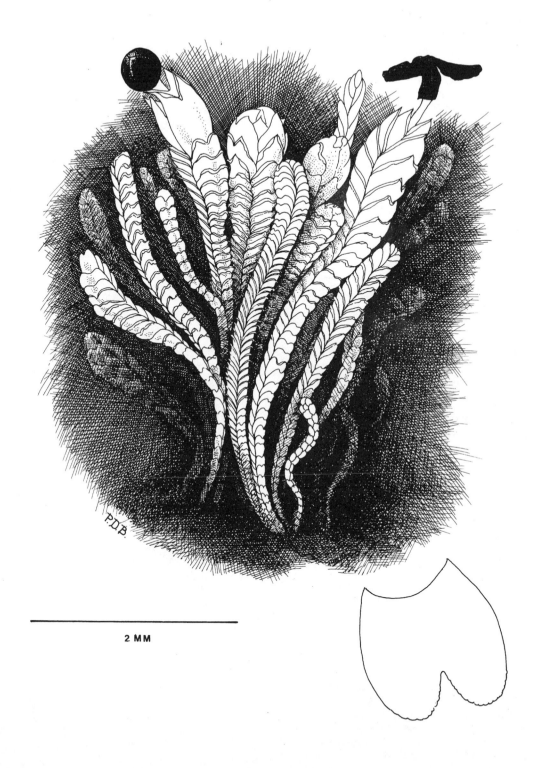

Gymnomitrion obtusum from British Columbian specimen. A leaf is shown lower right.

Gymnomitrion Corda

Name
Means "naked cap," in reference to the usual absence of a perianth.

Species
Six species in the region.

Habit
Forming white to olive-colored or brown turfs or mats of stiff, erect to reclining shoots. The shoots have the leaves very closely overlapping, making them appear wormlike.

Habitat
On exposed to somewhat sheltered rock surfaces from near sea level to alpine elevations, also in well-drained tundra and among boulders of boulder slopes.

Reproduction
Sporophytes occasional, usually not conspicuous, barely emergent on short seta, maturing in spring to summer.

Local Distribution
Widespread at low elevations near the coast and in mountains, from Alaska southward to California.

World Distribution
A predominantly Northern Hemisphere genus of mountains, but some species also in mountains of the Southern Hemisphere.

Distinguishing Characteristics
The wormlike rigid shoots that are usually condensed into tufts or short turfs and the usually whitish color, making many species resemble lichens at a quick glance, are characteristic.

Similar Genera
Some species of *Marsupella* are of similar size, but most are not wormlike. *Anthelia* is often whitish, but the color is the result of powdery covering or fungal threads, and all species of *Anthelia* occur in wet sites, while *Gymnomitrion* is white through scarcity of chlorophyll. *Pleurocladula* is also in wet sites and the leaves are pale green, but the leaves are concave and relatively widely spaced compared to those of *Gymnomitrion*.

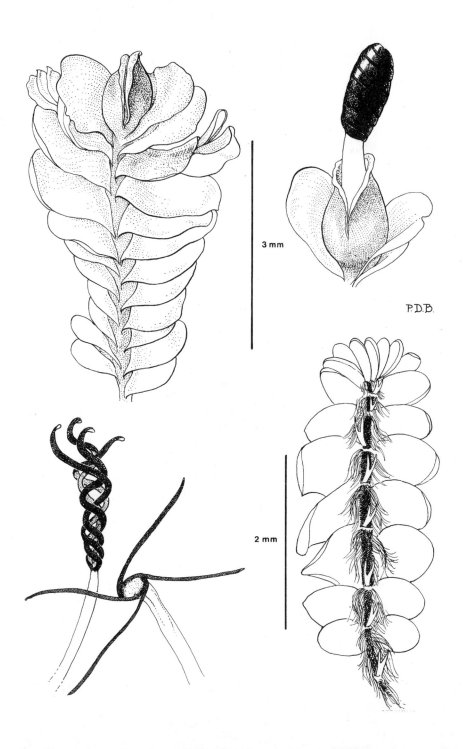

Gyrothyra underwoodiana from British Columbian specimen. Lower left shows an opened sporangium and one that has been rewetted. Lower right shows the ventral view of the underleaves and purple patches on the stem. Upper right shows the sporangium with its unusual spiral lines of opening.

Gyrothyra M. A. Howe

Name
Means literally "twisted door," in reference to the helically twisted openings of the elongate sporangium.

Species
One species: *G. underwoodiana*.

Habit
Forming pale green to sometimes pinkish mats or turfs of creeping to suberect shoots firmly affixed to the substratum.

Habitat
Disturbed mineral soil of open to somewhat shaded sites, generally at lower elevations. Pinkish plants appear in exposed areas, while in more shaded sites the plants are pale green.

Reproduction
Sporophytes abundant in spring. The unusual spiral dehiscence lines makes the opening sporangium an intriguing phenomenon to observe; as the four parts uncoil, the elaters and spores are thrown out.

Local and World Distributions
Endemic to western North America. Mainly along the coast, from southeastern Alaska to the northern half of California.

Distinguishing Characteristics
The elliptical purplish pads associated with the tufts of rhizoids on the undersides of the stem, coupled with the unlobed lateral leaves, are very distinctive. The helically opening divisions of the elongate sporangium are shared only with *Calypogeia* in the local flora.

Similar Genera
Chiloscyphus is of similar size and form, but lacks the purplish pads and clusters of rhizoids; sporangia of *Chiloscyphus* have longitudinal lines of opening. *Nardia* plants also lack the purplish pads, and sporangia are spherical. *Chiloscyphus* similarly lacks the purplish pads.

Comments
This unusual genus is placed in its own family, Gyrothyraceae. Its abundance has possibly been enhanced by human activity, since it invades disturbed mineral soil sites. It is often abundant in recently logged-over areas.

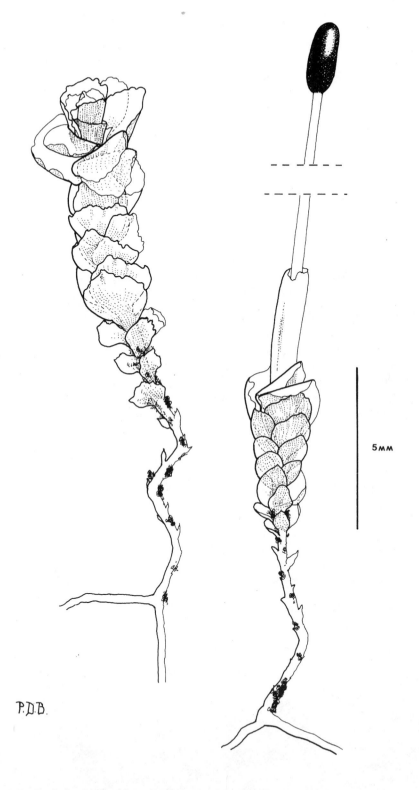

Haplomitrium hookeri from Washington State specimen.

Haplomitrium Nees

Name
Refers to the stem calyptra that serves as a protective sleeve for the developing sporophyte; means literally "simple cap."

Species
One species in the region: *H. hookeri*.

Habit
Short erect pale green plants with lowermost colorless rootlike portion deeply embedded in substratum; sometimes forming dense colonies. The living plants somewhat resemble miniature brussels sprouts.

Habitat
Humid, usually somewhat shaded, late-snow areas near or above tree line in mountains.

Reproduction
Sporophytes frequent, maturing in late summer; the seta is pale greenish when immature and is sheathed at the base by an elongate white stem-calyptra.

Local Distribution
Rare; mainly in near-coastal mountains, from southeastern Alaska to Washington.

World Distribution
The genus shows a very broken distribution in both hemispheres; it is predominantly in milder temperate and humid climatic zones.

Distinguishing Characteristics
The short erect fleshy shoots that arise from a colorless subterranean system in which the leafy shoots look like miniature brussels sprouts, plus the late-snow-bed habitat are distinctive characteristics. The pale green seta, when young, is also a common feature, as is the unusually elongate stem-calyptra (often more than 5 mm).

Similar Genera
Some species of *Lophozia* occupy similar habitats, but the leaves are always lobed, while those of *Haplomitrium* are unlobed. *Schofieldia* also has lobed leaves and occurs in habitats similar to those of the *Haplomitrium*, but the tight turfs of plants that lack rootlike organs immediately separate it.

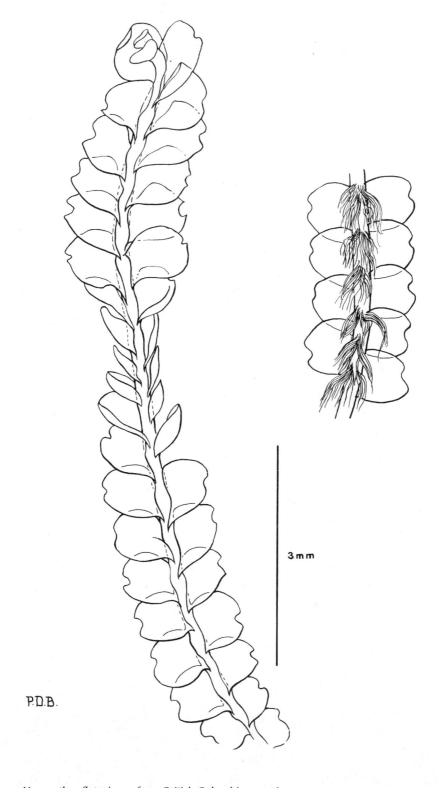

Harpanthus flotovianus from British Columbian specimen.

Harpanthus Nees

Name
Describes the curved pouchlike structure that contains the female sex organs and from which the sporophyte emerges.

Species
One species in the region: *H. flotovianus*.

Habit
Pale green to yellowish green or brownish, irregularly branched, suberect to creeping shoots loosely affixed to the substratum by rhizoids, usually forming loose soft tufts 2 to 4 cm tall.

Habitat
Terrestrial on wet to damp humus or silt of banks of streams and water bodies, chiefly at subalpine elevations; frequently intermixed with other bryophytes.

Reproduction
Sporophytes not observed in local material, but the species is sufficiently widely distributed to suggest that they occur. In European material the perianths appear lateral when sporangia emerge. The sporangium is ovoid, and the seta is relatively short, so that the sporangia often are buried among the tufts of leafy shoots.

Local Distribution
On mountains from Alaska to Washington.

World Distribution
Widespread in arctic and mountainous regions of the Northern Hemisphere. It appears to be most frequent on western sides of continents and rare elsewhere.

Distinguishing Characteristics
The obliquely oriented lateral leaves with strongly decurrent bases, the shallow sinus of the asymmetric leaves, and the relatively simple small underleaves are characteristic features. The subalpine moist habitat is also common in conjunction with these features.

Similar Genera
Some species of *Lophozia* somewhat resemble *Harpanthus*, but most have a deeper sinus in the bilobed leaves, and plants tend to be more rigid than *Harpanthus*. The restriction to higher elevations or latitudes also reduces the number of similar species of *Lophozia* that can be confused with it. *Chiloscyphus* is usually in wetter habitats, and careful examination reveals small bilobed underleaves in this genus, absent in *Harpanthus*. *Lophocolea* is also similar, but its tiny underleaves are bilobed, and the lateral leaves frequently have sharp points to the two lobes.

Herbertus aduncus from British Columbian specimen.
Lower left shows ventral view of the underleaves; upper right shows an opened sporangium.

Herbertus S. Gray

Name
In honor of T. Herbert, a nineteenth-century patron of the sciences.

Species
Three species, possibly four, in the region.

Habit
Plants varying from irregularly branched to unbranched or pinnately branched with branches widely spaced, some plants with frequent slender rootlike branches arising perpendicular from the undersurface of the shoot; color varying from nearly black to wine-red or rusty brown. Size varying from leafy shoots as little as 1 to 1.5 mm in diameter to as wide as 4 to 5 mm. Plants usually forming erect to suberect shoots more or less perpendicular to the substratum, occasionally creeping. Shoots vary in length from 2 to 3 cm. to as much as 30 cm.

Habitat
Extremely diverse: as epiphytes forming large rounded masses on trunks and branches of trees; on rock surfaces forming turfs or rounded masses; when terrestrial, especially in peatland, tundralike habitats, or fog forests, forming deep or shallow turfs of erect plants.

Reproduction
Generally vegetative, probably dispersed by fragmentation of brittle plants. Sporophytes known only in epiphytic plants of *H. aduncus*, appearing in late spring to summer.

Local Distribution
Most species confined to near the coast from sea level to subalpine sites from Alaska to northern Oregon.

World Distribution
A predominantly tropical and subtropical genus, in North America reaching its greatest diversity on the humid Pacific Coast; a genus of mainly very wet climatic regions.

Distinguishing Characteristics
The rusty brown to reddish brown color associated with the deeply bilobed leaves that give the shoots a mosslike appearance make this a very distinct genus.

Similar Genera
Tetralophozia is similar in color, and sometimes in appearance of the tufts, but the leaves are obviously four-lobed (not two-lobed, as in *Herbertus*) when viewed with a hand lens.

Comments
H. aduncus is the most widespread species in the genus, from northern Alaska to northern Oregon. The others are mainly near the coast of British Columbia and adjacent Alaska.

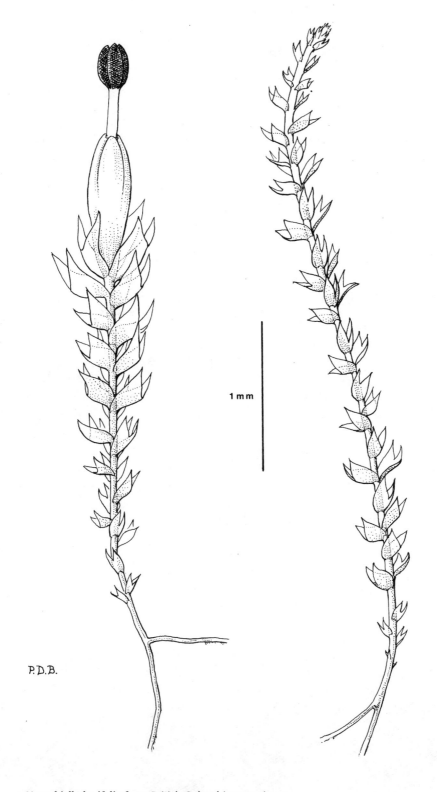

Hygrobiella laxifolia from British Columbian specimen.

Hygrobiella Spruce

Name
Refers to the wet habitat in which the plant grows, and means literally "little wet one."

Species
One species in the region: *H. laxifolia*.

Habit
Forming dark green to nearly black dense short turfs of irregularly branched shoots; plants usually 1 to 3 cm tall.

Habitat
Predominantly in rock crevices or depressions near streams, usually irrigated by seepage, in the splash zone or within the spray zone of waterfalls.

Reproduction
Sporophytes maturing in spring to summer; occasional.

Local Distribution
Not well documented; most collections taken from near the coast, from sea level to subalpine elevations from Alaska to Oregon.

World Distribution
Near-coastal and in mountains in eastern and western North America; also in Atlantic and alpine Europe and near-coastal East Asia.

Distinguishing Characteristics
The slender, shallow, turf-forming dark green to nearly black small plants, the sharply two-lobed leaves, and the humid near-stream habitat are useful characteristics that mark this plant.

Similar Genera
Although *Cephalozia* is similar in size, the plants are never dark green to nearly black; *Cephaloziella* is of similar size and often alike in color, but is rarely in sites where *Hygrobiella* is found. *Anthelia*, of similar form, usually has powdery bluish green shoots (*A. julacea* may be troublesome, but check for the depth of the sinus between the lobes; in *Anthelia* it is at least one half the leaf length, but in *Hygrobiella* it tends to be less than one third the leaf length). *Gymnocolea* is sometimes similar in size, but the leaves are nearly round in outline, rather than elongate, as in *Hygrobiella*. *Sphenolobopsis* and *Eremonotus* are also very slender plants on shaded rock faces, but not associated with water-spray from streams.

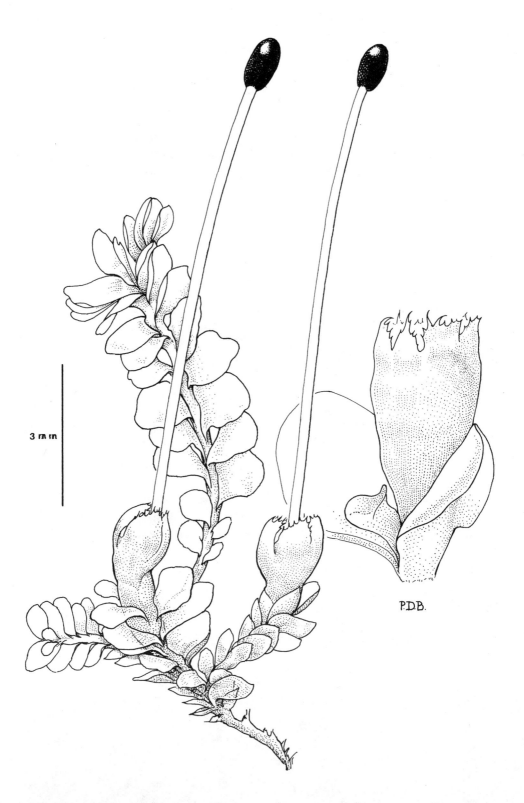

Jamesoniella autumnalis from British Columbian specimen. Right shows a characteristic perianth.

Jamesoniella (Spruce) Carring.

Name
Refers to the fancied resemblance to the fern *Jamesonia*, named in honor of William Jamieson, a nineteenth-century botanist of Ecuador.

Species
Two species in the region.

Habit
Forms dense turfs of creeping to slightly suberect shoots varying from dark green to reddish purple, usually firmly attached to substratum.

Habitat
Mainly on decaying logs in open forest; occasionally on rock surfaces and in tundra.

Reproduction
Sporophytes maturing in late winter to early spring.

Local Distribution
Probably widely distributed in humid coniferous forests in the region from Alaska to California, but lacking in drier forest types and rare in tundra.

World Distribution
Circumpolar in the Northern Hemisphere, extending southward to tropical latitudes in Asia. Mainly a Southern Hemisphere genus.

Distinguishing Characteristics
The often reddish purple plants, the sporophytes produced in late winter and spring, the perianth with long cilia at the mouth, and the unlobed leaves are very distinctive features (especially for the widespread species, *J. autumnalis*).

Similar Genera
Plants of *Mylia taylorii* can be bright red-brown, but the plants are usually suberect and two to three times larger than *Jamesoniella*, and perianths are rarely present. *Jungermannia* has perianths that lack cilia, as does *Nardia*. *Gyrothyra* is also much larger, and has purple patches on the undersides of the stems associated with clumps of rhizoids; these features are lacking in *Jamesoniella*. *Plagiochila porelloides*, when lacking marginal teeth on the leaves, may resemble *Jamesoniella*, but the leaves of *Plagiochila* are usually strongly decurrent on the upper surface of the stem, while those of *Jamesoniella* are usually not; the flattened perianth of *Plagiochila* contrasts with the not flattened perianths, when mature, of *Jamesoniella*.

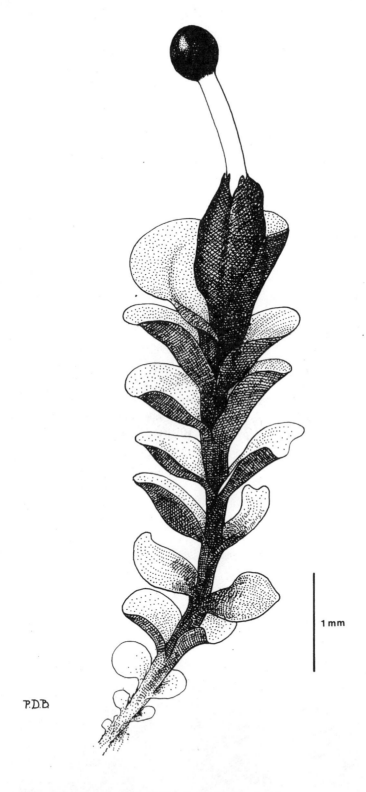

Jungermannia rubra from British Columbian material.

Jungermannia L.

Name
In honor of L. Jungermann, a seventeenth-century German botanist.

Species
At least fifteen species in the region.

Habit
Some species form short turfs, while others creep over the substratum; most species have plants that are 1 to 3 mm in diameter, but some exceed 5 mm. Color varies from pale to dark green to nearly black or pigmented with red. Plants are unbranched or irregularly branched.

Habitat
Diverse; some species are on bare mineral soil, others are on humid rock surfaces, and still others grow on decaying wood or tree trunks; some are submerged aquatics, usually on rocks in streams, and accumulate sand; occurring from sea level to alpine elevations.

Reproduction
Sporangia are spherical and dark brown to black when mature; some species produce sporophytes in spring, while others produce them during the summer; plants are brittle, thus dissemination can occur through fragmentation.

Local Distribution
Widespread throughout the region at all elevations from Alaska to California.

World Distribution
Widely distributed throughout the world.

Distinguishing Characteristics
The unlobed leaves and inflated puckered-mouth or beaked perianth are useful characteristics. Most species are dark green and relatively small (leaves often less than 1 mm in diameter) and creep over the substratum.

Similar Genera
In *Jamesoniella* the toothed mouth of the perianth is enough to separate it from *Jungermannia*. In *Gyrothyra* and *Chiloscyphus* the presence of small lobed underleaves separate them from most species of *Jungermannia*. *Arnellia* has nearly opposite leaves fused at the base, while those of *Jungermannia* are alternately arranged and not fused. *Odontoschisma* has leaves with clearly concave upper surfaces, facing the stem apex, while in *Jungermannia* leaves are not strongly concave. In *Nardia* and *Mylia* underleaves are present, and the colonies are usually formed of dense erect shoots, rather than creeping shoots as in most species of *Jungermannia*. Unfortunately, microscopic features are often needed to distinguish some species of *Jungermannia* from some of these genera. See also notes under *Cryptocolea*.

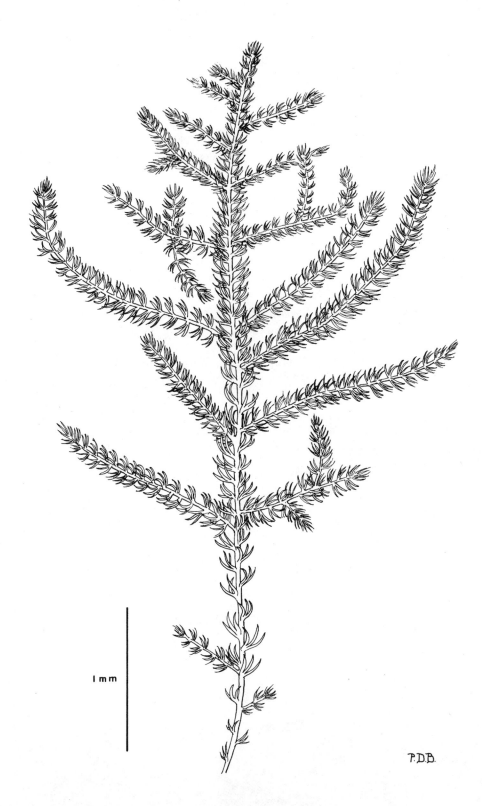

Kurzia pauciflora from British Columbian specimen.

Kurzia von Martens

Name
In honor of W. S. Kurz (1834–78), a German botanist who made many collections in the Indo-Malaysian region, and who, for a time, was curator of the Royal Herbarium in Calcutta.

Species
Three species in the region.

Habit
Forms dense turfs of dark brown to dark green filamentous, regularly branched shoots, appearing almost leafless unless viewed with a hand lens.

Habitat
Usually shaded sites in humid environments, especially on humic substrata: peat banks, rotting logs, and stumps.

Reproduction
Sporophytes infrequent in local material. The shoots are readily fragmented, and may serve to propagate the plants.

Local Distribution
Predominantly near the coast and at lower elevations from Alaska to northern California.

World Distribution
Widely distributed in humid climate areas, mainly near the coast or in mountains; most species are confined to the Southern Hemisphere.

Distinguishing Characteristics
The local species are all very slender, generally brownish, bear very tiny, deeply three-lobed leaves, are often pinnately branched, and form loose mats of interwoven shoots on a humic substratum. These features are usually enough to separate this genus.

Similar Genera
The brown or dark green color and very deeply three- to four-lobed leaves are usually enough to separate them from *Lepidozia* (in *Lepidozia*, plants are pale to rich green, never brownish). The regularly four-lobed leaves with large teeth on the lobe margins of *Tetralophozia* make it readily separable from *Kurzia*. Some specimens of *Herbertus* can be very small, but the leaves are deeply bilobed, not three- to four-lobed. *Blepharostoma* is rich green to pale green, has deeply divided leaves, and lobes of a single linear series of cells rather than two cells wide, as in *Kurzia*. *Blepharostoma* is never brownish in color. *Takakia* can be similar in size to *Kurzia*, but plants are always bright green, leaves are not lobed, and a complex rhizomatous system is present, which is never so in *Kurzia*. Of similar size and color are *Cephaloziella*, *Hygrobiella*, and *Sphenolobopsis*, but all of these genera have bilobed, not trilobed, leaves, and none have pinnate branching.

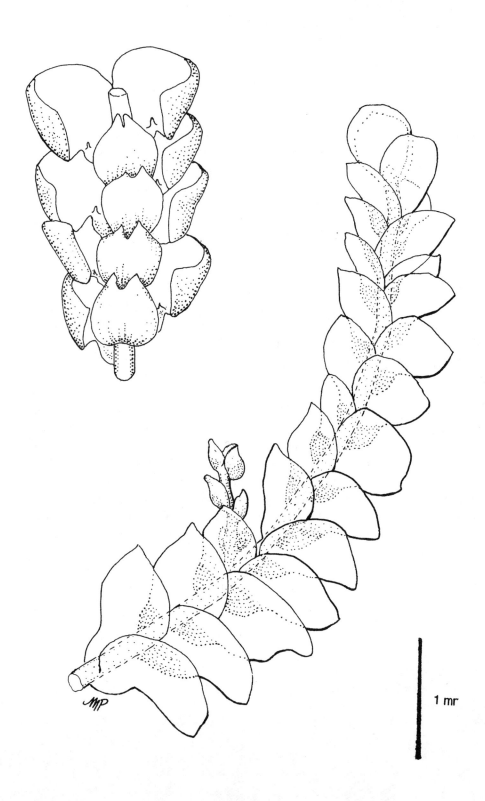

Lejeunea alaskana from Alaskan specimen. Ventral view on upper left.

Lejeunea Lib.

Name
In honor of nineteenth-century Belgian botanist L. S. Lejeune.

Species
A single species in the area: *L. alaskana*.

Habit
Tiny pale green to yellow-green strands forming pure tufts or growing among other bryophytes. The leafy shoots are unbranched or irregularly and little branched, usually less than 1 mm wide, and can reach lengths of up to 2 cm.

Habitat
On sedge tussocks in moist to wet somewhat calcareous soils where moisture is abundantly available in tundra.

Reproduction
Sporophytes unknown in the local species; the plant presumably disseminates by fragments of leafy stem.

Local Distribution
Arctic Alaska, where apparently confined to the north slope of the Brooks Range and Nahanni National Park, Mackenzie District, Northwest Territories, Canada.

World Distribution
The genus is widely distributed and diverse in the tropics and subtropics, extending northward mainly to warmer temperate latitudes. In eastern North America some species extend northward to Newfoundland.

Distinguishing Characteristics
The tiny, pale green plants in which the lateral leaves have a swollen pouchlike base, the relatively unbranched shoots, and the equally bilobed underleaf with a narrow sinus serve as useful identifying features.

Similar Genera
Radula is similar in size and color, but the plants of *Radula* are often regularly branched, flattened, and lack underleaves. *Calypogeia* is superficially similar, but it is never yellow green, the sinus of the underleaf is either lacking or wide, and the plants are decidedly flattened. *Metacalypogeia* closely resembles this genus, but in *Metacalypogeia* the leaves lack the bulging base, and lateral branches tend to be frequent; *Bazzania* has the same distinguishing features as *Metacalypogeia*. *Cololejeunea* is similar in size, but plants tend to be flattened over the substratum, have no suggestion of glossiness when dry (*Lejeunea* is somewhat shiny), and plants tend to have frequent lateral branches. *Cryptocolea*, of similar habitats, lacks underleaves, and the lateral leaves lack a bulging base.

Comments
This genus represents, in Alaska, a survivor of an extremely ancient flora that has persisted in areas that escaped the major recent glaciations. The species is related to those of Southeast Asia of Palaeotropical affinity.

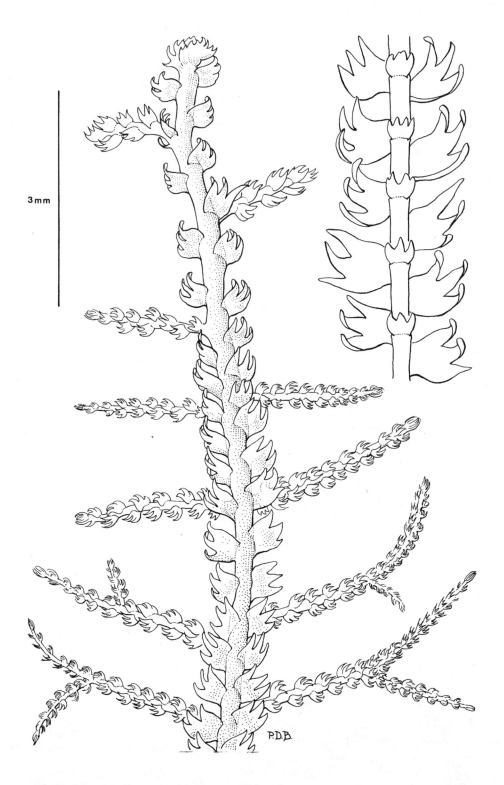

Lepidozia reptans from British Columbian specimen.
Upper right shows ventral view of underleaves.

Lepidozia (Dum.) Dum.

Name
Means "scaly branchlets," in reference to the scalelike leaves.

Species
Three species in the region.

Habit
Plants usually regularly pinnate, dark green to pale yellow green, varying in size with leafy shoots less than 0.5 mm in diameter to more than 2 mm; plants from less than 1 cm long to more than 6 cm; forming loose to dense turfs or carpets weakly affixed to substratum.

Habitat
Most frequent in shaded sites, on rotten logs or stumps, but some species on moist soil, especially near watercourses; from sea level to subalpine.

Reproduction
Sporophytes maturing from spring to summer, not frequent; the brittle plants probably important in vegetative propagation.

Local Distribution
Only *L. reptans* is widespread in coniferous forest, sometimes into tundra, in the region from Alaska to California. The other species are confined to near the coast from sea level to subalpine elevations in regions of highest precipitation of Alaska and British Columbia.

World Distribution
The genus is widely distributed in the world, reaching its greatest abundance and diversity in foggy forests of the tropics and subtropics.

Distinguishing Characteristics
The often pale green color, the three to four fingerlike lobes of the tiny leaves, the usual pinnate branching of the plants, and the frequent presence of rootlike branches emerging from the undersurface of the main shoots are useful characters. *L. sandvicensis* has such tiny leaves that it seems leafless.

Similar Genera
Although *Tetralophozia*, *Chandonanthus*, and *Blepharostoma* have deep fingerlike lobes of the leaves, the plants are never pinnate, as in *Lepidozia*; none of these has the rootlike branches that arise from the undersurface of the stem in *Lepidozia*. *Kurzia* is often pinnate and brownish, and the leaf lobes are extremely slender compared to *Lepidozia*; it also lacks rootlike branches that emerge from the underside of the stem.

Comments
Sometimes called "Mickey Mouse Hands" based on the frequently three-lobed leaves, comparable to the three fingers of Mickey Mouse.

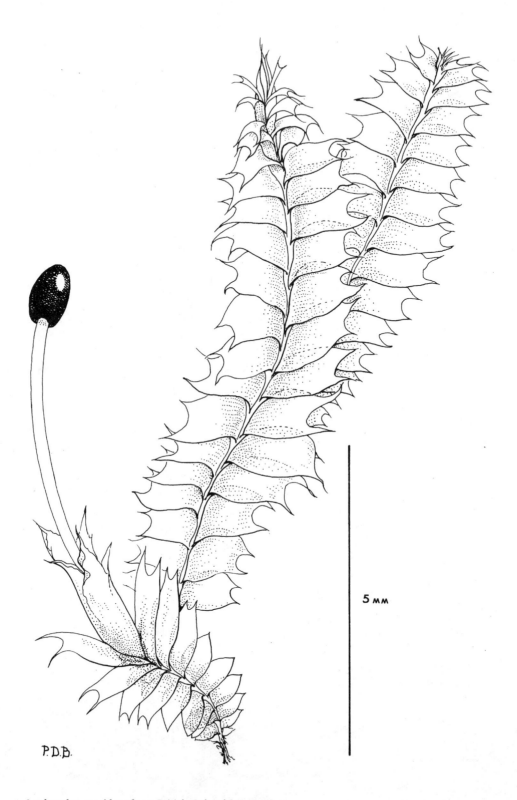

Lophocolea cuspidata from British Columbian specimen.

Lophocolea (Dum.) Dum.

Name
Describes the toothed ridges of the sleevelike perianth of some species.

Species
Three species in the region.

Habit
Creeping, somewhat flattened, irregularly branched shoots varying from dark green to pale yellow green, firmly affixed to substratum in most cases; plants occasionally forming loose carpets, but usually flattened to substratum.

Habitat
Epiphytic on living trees, cliffs, rotten logs, among herbs, and on mineral soil, from shaded to relatively open sites within forests to moist disturbed banks.

Reproduction
Sporophytes frequent, maturing in late autumn to spring; gemmae frequent in *L. minor*, bright yellow green.

Local Distribution
Two species frequent near the coast where precipitation is relatively high, from Alaska to California, but *L. minor* more common in interior regions.

World Distribution
Widely distributed in temperate and humid tropical areas, reaching its greatest diversity in the tropics and Southern Hemisphere.

Distinguishing Characteristics
The three local species are mainly pale yellow green in color and have bilobed lateral leaves; plants creep along on the substratum and are irregularly branched. Plants growing on tree trunks, however, tend to be darker green. These features, plus the sporophytes that are produced in late winter or early spring, earlier than any other hepatic that resembles it, provide useful characteristics.

Similar Genera
Some species of *Lophozia* that bear bilobed leaves are superficially similar to *Lophocolea*, but plants tend to be suberect, and sporophytes are produced in summer. *Geocalyx* is similar in color and habitat to *Lophocolea*, but sporophytes bear elongate rather than spherical sporangia, and the sporophytes emerge from a subterranean pouch rather than an exposed perianth. *Chiloscyphus* is also similar, but leaves tend to be blunt-lobed, and the plants occupy wet sites rather than the well-drained habitats of *Lophocolea*. *Fossombronia longiseta* also produces sporophytes in the early spring, but the sporophytes emerge from a protective sleeve that arises dorsally on the thallus far back from the apex, and the sporangium opens irregularly (in *Lophocolea* the perianth terminates a shoot, and the sporangium opens by longitudinal lines); the "leaves" of *Fossombronia* lack lobes. *Gyrothyra* is superficially similar, but the purple patches on the undersurfaces of the shoots, usual in *Gyrothyra*, are lacking in *Lophocolea*, and lateral leaves are unlobed. See also *Acrobolbus* and *Harpanthus*.

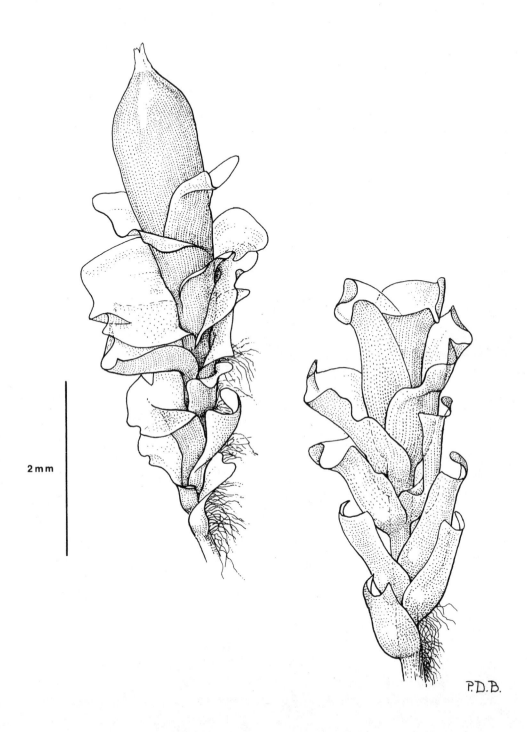

Lophozia gillmanii from British Columbian specimen.

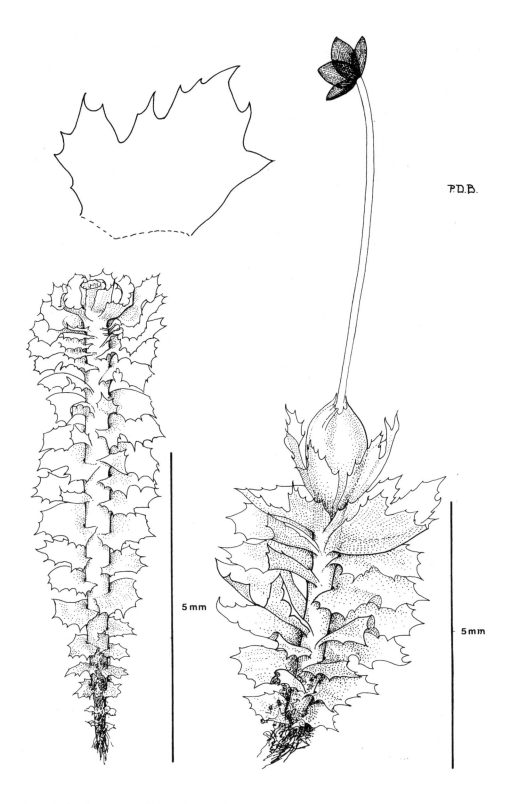

Lophozia incisa from British Columbian specimen.

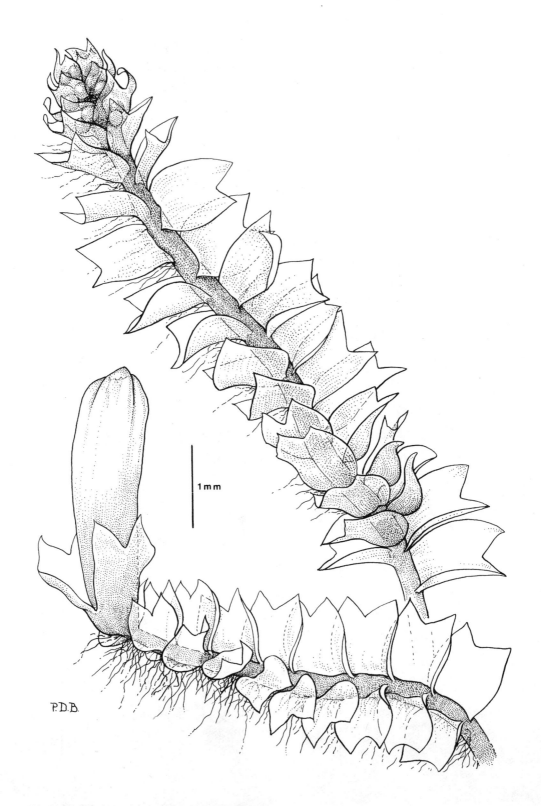

Lophozia ventricosa from British Columbian specimen.

Lophozia (Dum.) Dum.

Name
Refers to the sharp points on the leaf lobes of many species, in contrast to the genus *Jungermannia*, which lacks such points.

Species
At least twenty-nine species in the region.

Habit
Diverse; most plants are pale to dark green, but some are decidedly tinged with reddish purple or brown; they sometimes form creeping carpets of interwoven plants, and at other times form tufts or turfs of erect to suberect shoots; plants vary from 1 to 2 cm long and near 1 mm in diameter to more than 6 cm long and 5 mm in diameter; branching is irregular, and some species are often unbranched.

Habitat
Diverse, mainly terrestrial, but also epiphytic; on rotten logs and on rock, generally in humid sites, both in forests and open areas, from near sea level to alpine elevations.

Reproduction
Sporophytes frequent in some species, rare in others, maturing in spring to autumn; gemmae frequent in many species, of distinctive size and color dependent on the species, the detail visible only with microscopic examination.

Local Distribution
The genus is widespread in the region, reaching its greatest diversity in Alaska, with lowest diversity in California.

World Distribution
A predominantly Northern Hemisphere genus and widespread in temperate and arctic regions.

Distinguishing Characteristics
The genus is a large and complex one in the region, and the features that distinguish species are mainly visible only with a compound microscope. The genus also shows considerable variability both within and among species, which further adds to the confusion. Most species have bilobed leaves, but others have three or four lobes. Even in a single shoot, leaves of a variable number of lobes can occur. Many species form turfs of erect to suberect rather than creeping shoots, and leaves appear somewhat fleshy. Color is extremely variable, but most species are light green in color; some however, are bluish green, others have a tint of purple, and still others are somewhat brownish. With experience, it is possible to recognize the genus, but the concept is usually based on familiarity with the species, gained through microscopic examination and study.

Note
This is an extremely conservative concept of the genus. It is sometimes divided into several separate genera that are accepted by many researchers.

Similar Genera
See notes under the following genera: *Anastrepta, Anastrophyllum, Barbilophozia, Chandonanthus, Harpanthus, Lophocolea, Schofieldia, Tetralophozia,* and *Tritomaria.*

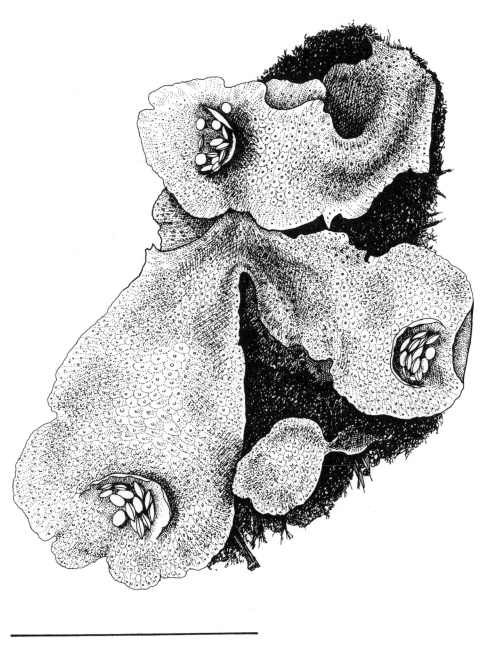

Lunularia cruciata from British Columbian specimen.

Lunularia Adans.

Name
Refers to the crescent-shaped dorsal pocket that contains gemmae.

Species
A single species: *L. cruciata*.

Habit
Thallus pale, somewhat glossy green, with surface dotted with tiny pores and marked by cell-like pentagonal boundaries of the chambers beneath; usually depressed against the substratum and affixed by colorless rhizoids.

Habitat
On somewhat shaded soil of gardens and a very common greenhouse weed. Also in shaded humid sites along watercourses on soil and rocks, especially in California.

Reproduction
Gemma production abundant throughout the growing season; sporophytes infrequent in much of its local range. The species appears to have been widely disseminated with horticultural plants.

Local Distribution
From southwestern portion of British Columbia where restricted to urban gardens and greenhouses, southward in natural habitats along streams, especially in California.

World Distribution
Widely distributed in temperate to warmer climates, possibly the range expanded through its dissemination as a greenhouse weed. It is native to Europe, Africa, South America, Australia, and probably elsewhere.

Distinguishing Characteristics
The crescent-shaped gemma cups are unknown in any other thallose liverwort; the thallus is also a somewhat shiny yellow green.

Similar Genera
Marchantia, the only other genus that produces gemma cups locally, has the cups with a circular outline and toothed margin; the thallus is also usually dark green, rather than yellow green.

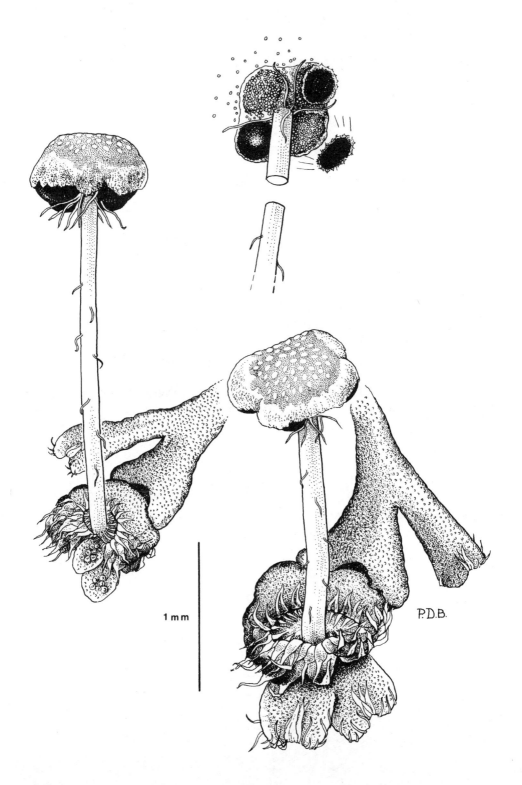

Mannia fragrans from British Columbian specimen. Upper right shows a sporangium releasing its lid.

Mannia Opiz.

Name
In honor of a person of the surname Mann, about whom the author of the genus unfortunately provided no details.

Species
Three species in the region.

Habit
Pale to dark green thalli with undersurface, especially near margins, dark purple to nearly black with overlapping scales. Margins curling upward and inward when dry make the elongate thalli narrow and tubelike in appearance. The stalk of the receptacle is generally purplish and slender.

Habitat
On mineral soil, usually near the shelter of other plants, in semiarid areas and on open slopes and terraced outcrops.

Reproduction
Sporophytes maturing in early summer, erratically produced, dependent on favorable periods of wet weather. Thalli are brittle when dry, and fragments could serve as vegetative propagants.

Local Distribution
Poorly documented, but in British Columbia scattered in the semiarid interior in steppe habitats and more widespread southward to California.

World Distribution
Widely scattered in the Northern Hemisphere and also scattered in the Southern Hemisphere, rarely common, and predominantly in sites that dry out during part of the year.

Distinguishing Characteristics
The narrow thallus that has the margins strongly incurved when dry, making the plants tubelike in appearance, and the fringe of white scales at the base of the stalk of the receptacle plus the sporangium that opens by a lid, rather than longitudinal lines, make this very distinctive.

Similar Genera
Targionia also has thalli that curl up when dry, but the sporophytes that terminate the undersurface of the lobes make this genus very distinctive. *Asterella* has no scales at the base of the receptacle, and the white involucres around the sporangia are very distinctive. *Reboulia* thalli do not become tubelike when dry.

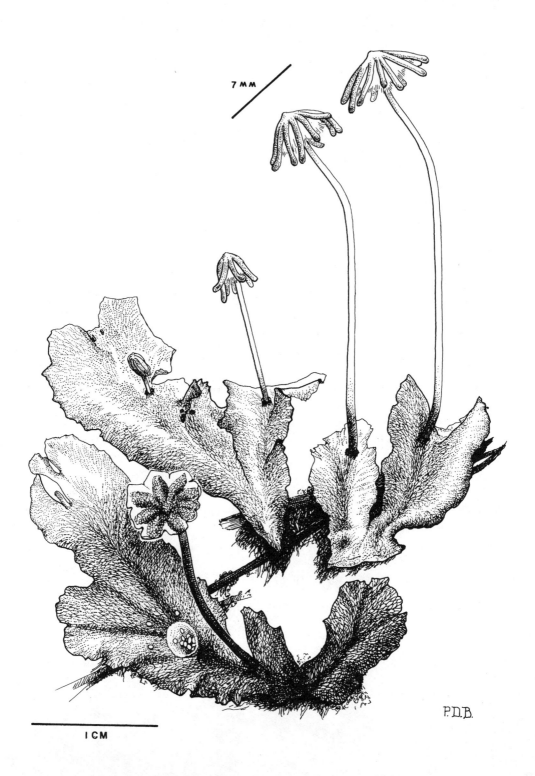

Marchantia polymorpha from British Columbian specimen.
The lower plant has an antheridium-producing receptacle and a gemma cup.

Marchantia L.

Name
In honor of N. Marchant, a seventeenth-century director of the gardens of Count Gaston of Orléans in Blois, France.

Species
One species in the region.

Habit
Thalli generally flattened and light green, showing outlines of air chambers and their pores visible without a hand lens. Thalli can be up to 5 cm long and 2 cm wide; they often form mats of overlapping lobes that are firmly affixed to the substratum by numerous rhizoids.

Habitat
Frequent as a garden and greenhouse weed, especially in somewhat shaded moist sites; also in ditches and seepy areas from sea level to subalpine and alpine elevations. At higher elevations it reaches its greatest abundance and luxuriance in moss mats margining snow-melt streamlets.

Reproduction
Sporophytes mature in summer to autumn, but receptacles appear in spring. The female receptacles tend to be pale green when young, yellowish brown when old; while male receptacles tend to have a purplish surface. Gemma cups are most frequent in autumn and spring, but are present throughout the year, especially on thalli without sexual receptacles.

Local Distribution
Widely distributed throughout the area from sea level to subalpine elevations.

World Distribution
The genus is widely distributed throughout the world.

Distinguishing Characteristics
When with receptacles, the plant is unmistakable; those bearing the sporangia are umbrella-like, with numerous elongate lobes; the rounded gemma cups with toothed rims are also distinctive. Thalli often have a central darkened band of cells along the thallus length, a feature absent in other local genera that resemble *Marchantia*.

Similar Genera
Lunularia also produces gemma cups, but cups are crescent shaped rather than circular in outline; thalli of *Lunularia* are yellow green and glossy, while those of *Marchantia* are a non-glossy light to dark green. The surface outline of the air chambers, especially in the center line of the thallus, shows them to be narrow and elongate in *Marchantia*, but broad and short in *Lunularia*. The female receptacle form differs from all other local genera.

Comments
Marchantia is very frequent on bonfire sites a year or two after the fire; the same is true for forest fire areas, where this liverwort can be very abundant. It is one of the few bryophytes that can become a pernicious greenhouse weed, undoubtedly spreading through the greenhouse by splashing the gemmae out of their cups during watering. Sometimes the high elevation populations are treated as another species, *M. alpina* (Nees) Burgeff, but the features separating it from *M. polymorpha* are extremely subtle.

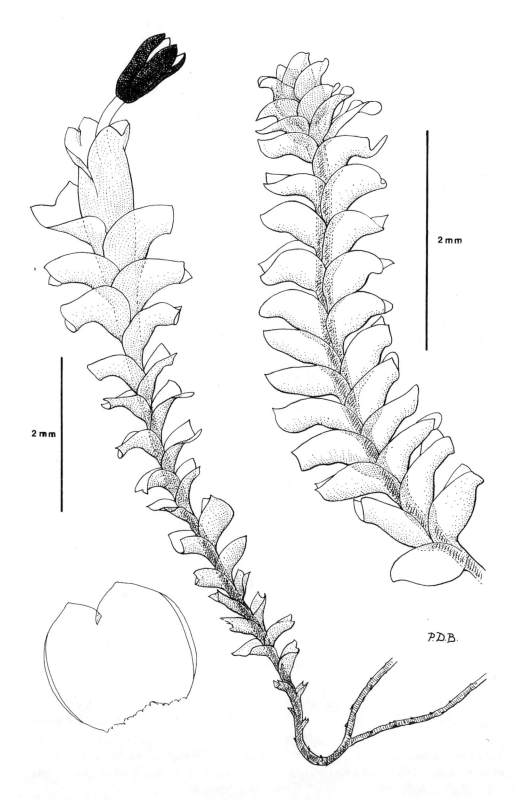

Marsupella emarginata from British Columbian specimen.

Marsupella Dum.

Name
Means "little pouch or purse," describing the enclosure of the developing sporophyte, the small perianth almost completely enveloped by perichaetial leaves.

Species
Fifteen species in the region.

Habit
Most species are turf-formers and vary in color from dark green, red brown, purplish brown, to nearly black; turfs can be less than 1 cm tall or, in some species, can be 5 cm or more tall.

Habitat
Either terrestrial on mineral soil or on rock surfaces, often in wet or moist environments, in open to somewhat shaded sites; frequent in sites where snow persists in alpine and subalpine regions and on rock near watercourses, extending to near sea level. Many species are submerged part of the year; a few are in well-drained tundra.

Reproduction
Sporophytes frequent in spring to summer in several species, not noted in others. The white seta and the spherical brown to nearly black sporangia contrast markedly with the dark leafy plants.

Local Distribution
The genus is widely distributed throughout the region from sea level to alpine elevations.

World Distribution
Predominantly in cooler portions of the Northern Hemisphere, but a few species occur in tropical latitudes and at higher latitudes of the Southern Hemisphere.

Distinguishing Characteristics
The generally dark color (red brown and black predominating) of the plants, the erect, irregularly branched or unbranched shoots that form turfs, the equally bilobed leaves that diverge outward from a somewhat sheathing base, the lack of underleaves, and the generally open sites where the genus occurs are all useful characteristics.

Similar Genera
Anastrophyllum can resemble *Marsupella* in form and color, but only *A. assimile* is likely to cause confusion, and the leaves of this species tend to curve with the points facing to one side of the shoot; in *Marsupella*, all leaf points face the apex of the shoot. *Gymnocolea* is similar in color but the inflated perianths (usually present) easily separate it from *Marsupella*, which lacks such perianths. *Gymnomitrion* species are very wormlike in appearance with the leaves strongly imbricated, so that some small species of *Marsupella* are difficult to distinguish from it. The same is true for *Hygrobiella*, *Eremonotus*, and *Sphenolobopsis*.

Note
The genus *Anomomarsupella* is sometimes segregated from *Marsupella*.

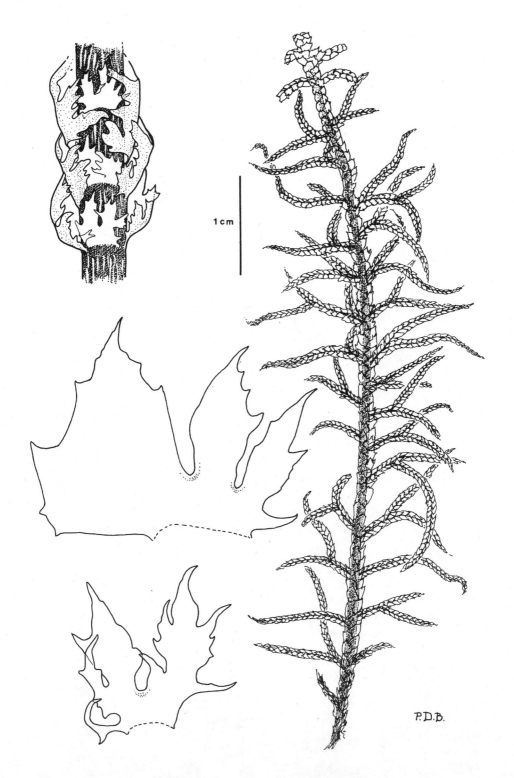

Mastigophora woodsii from British Columbian specimen. The underside of the shoot with underleaves is shown upper left, and a lateral leaf (large) and underleaf (small) are shown lower left.

Mastigophora Nees

Name
Means "bearing a whip," which describes the whiplike lateral branches.

Species
One species in the region: *M. woodsii*.

Habit
Forming tall turfs of loosely interwoven, often arching, regularly pinnate plants up to 6 cm in height, bright golden to orange brown, loosely affixed to substratum.

Habitat
Near watercourses, often within spray of cascades or waterfalls in extremely high precipitation regions.

Reproduction
Sporophytes and gemmae unknown in the region; possibly fragmentation of the plants serves in local dispersal.

Local Distribution
Extremely rare; known from the Queen Charlotte Islands and Pitt Island, British Columbia.

World Distribution
The genus is widely distributed in the humid tropics and subtropics, especially at higher elevations. In the Northern Hemisphere, only *M. woodsii* occurs in cooler temperate areas and shows a very interrupted distribution in oceanic climates of British Columbia, the Faeroe Islands, Great Britain, Taiwan, and humid climates on the flanks of the Himalayas.

Distinguishing Characteristics
The regularly pinnate bright brownish orange to reddish orange plants that form loose mats and bear many-lobed, large-toothed leaves are extremely distinctive.

Similar Genera
Only *Ptilidium ciliare* is likely to be confused with *Mastigophora*, but in *Mastigophora* the lateral branches are tapered to the tips, while in *Ptilidium* they are not. In *Ptilidium* the leaves bear closely placed cilia on the margins, and in *Mastigophora* the margins have widely spaced elongate teeth. The plants of *Ptilidium* tend to be soft, while those of *Mastigophora* tend to be rigid.

Comments
This genus probably is a surviving remnant of a very ancient flora that existed in the region millions of years ago; its reproductive inefficiency has restricted its wider dissemination.

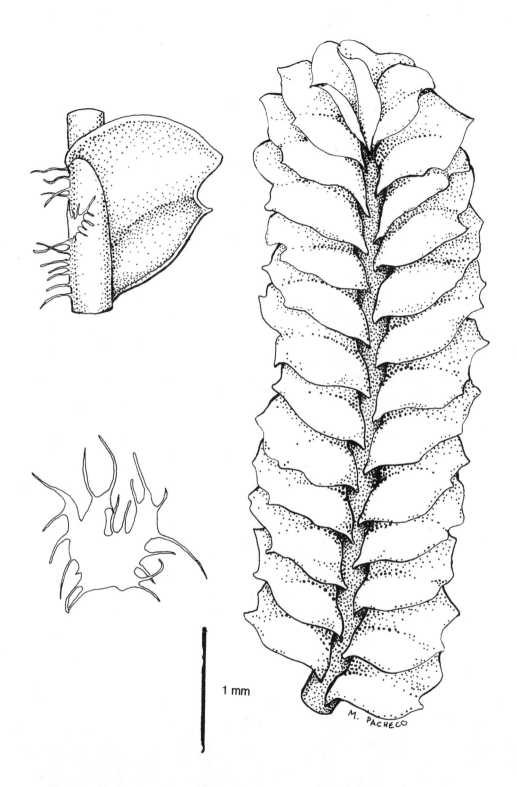

Mesoptychia sahlbergii from Alaskan specimen, augmented by reference to illustrations made by R. M. Schuster.

Mesoptychia (Lindb.) Evans

Name
Describes the perianth that possesses a high dorsal crest or fold.

Species
A single species: *M. sahlbergii*.

Habit
Creeping to suberect brownish-green to brownish purple robust plants, up to 4.5 mm wide (sometimes larger) and with stems 4 to 6 cm long, usually in loose mats but sometimes forming tufts. Rhizoids usually abundant, often purplish, especially at attachment to stem. Perianth well developed from a swollen base, with the mouth suddenly constricted to a short ciliate beak.

Habitat
In damp calcareous soil or peat or over limestone.

Reproduction
Sporangia infrequent, oblong to ellipsoid, emerging from a perianth at the apex of a subterranean pouch; sporangia maturing in summer.

Local Distribution
Confined, in the region, to arctic and boreal Alaska and Yukon.

World Distribution
Scattered in arctic and boreal North America and Greenland and in arctic Asia.

Distinguishing Characteristics
The asymmetrically, shallowly bilobed leaves, the ciliate bilobed underleaves, the purplish tint to rhizoids, and the usual red brown to nearly black color plus its northern distribution mark this genus.

Similar Genera
Similar in form to *Lophozia rutheana*, especially in the form of leaves; the underleaves of *Mesoptychia*, however, have numerous long regular cilia, while those of the *Lophozia* are few and irregular; *Mesoptychia* has purplish pigment in some rhizoids (especially near the base), and those of *Lophozia* are colorless. The dark pigmentation of *Mesoptychia* will separate it readily from superficially similar *Chiloscyphus*, *Lophocolea*, and *Harpanthus*. The *Harpanthus* has underleaves that are unlobed, while the other genera noted have bilobed inconspicuous underleaves.

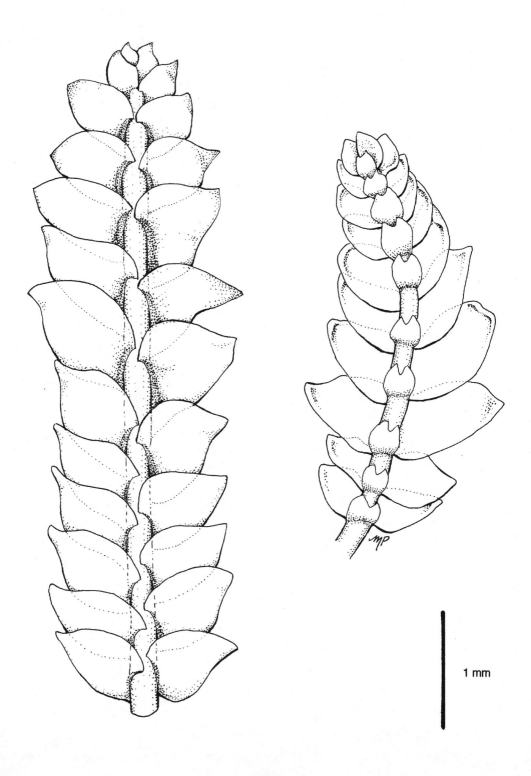

Metacalypogeia schusterana from Alaskan specimen. The right figure is a ventral view.

Metacalypogeia (Hatt.) H. Inoue

Name
Means "associated with *Calypogeia*," to denote its relationship to this genus.

Species
Two species in the region.

Habit
Pale green to dark green, sometimes yellowish green, creeping, unbranched or weakly branched, occurring as strands with other bryophytes or as pure tufts or thin mats.

Habitat
Moist humus on shaded calcareous cliffs; in cliff crevices and on shelves.

Reproduction
Sporophytes not noted in local material; plants possibly disseminated through fragmentation.

Local Distribution
Arctic Alaska and to Nahanni National Park in Mackenzie District, Northwest Territories.

World Distribution
The local species shows a broken distribution in arctic North America: northern Alaska, Attu Island in the Aleutian Chain, Bathurst Island, and also in the Gulf of St. Lawrence area in Quebec, western Newfoundland, and northern Nova Scotia (Cape Breton Island). The genus is also found in Southeast Asia.

Distinguishing Characteristics
The tiny pale green to dark green plants usually in pure colonies over rock or on humus of shaded cliff shelves and chimneys, the slightly notched tips of the lateral leaves and the bilobed underleaves, and its usually calcareous substratum serve as useful distinguishing features.

Similar Genera
It can usually be distinguished from *Bazzania* by the usual Y-shaped branching and the frequent rootlike branches in *Bazzania*, which are lacking in *Metacalypogeia*; furthermore, the shoots of *Metacalypogeia* are generally less than 1 mm wide, while those of *Bazzania* are usually wider. Some species of *Radula* superficially resemble *Metacalypogeia*, but underleaves are lacking in *Radula*. While *Lejeunea* has underleaves and is similar in size to *Metacalypogeia*, the lateral leaf of *Lejeunea* has a small swollen toothed lobe; this is absent in *Metacalypogeia*.

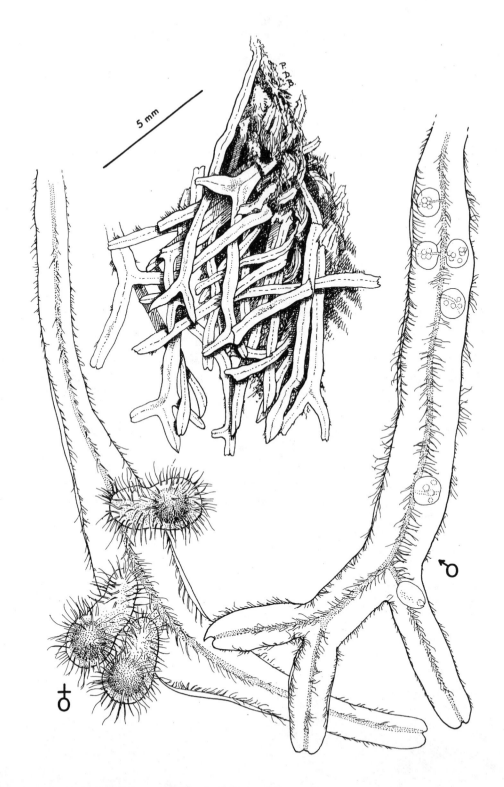

Metzgeria conjugata from British Columbian specimen.
Detailed drawings show the undersurfaces with sexual branches.

Metzgeria Raddi

Name
In honor of J. B. Metzger, a copper engraver and friend of the author of the genus.

Species
Three species in the region.

Habit
Plants pale green to yellow green, forking thalli, closely appressed to the substratum or forming loosely affixed mats of interwoven thalli all pointing in the same direction. The thallus midrib is very apparent with hand-lens examination.

Habitat
Epiphytic, epilithic, and occasionally terrestrial among other plants, especially on cliff shelves.

Reproduction
Sporophytes occasional in *M. conjugata*, unknown in the other local species; gemmae abundant in *M. temperata*, occupying margins of the apices of the thalli, less frequent in other species.

Local Distribution
The genus is widely distributed in the region from sea level to alpine elevations, reaching its greatest diversity in coastal portions of southeastern Alaska, adjacent British Columbia, and southward to California.

World Distribution
The genus is widely distributed throughout the world.

Distinguishing Characteristics
The small forked thalli that bear a very distinctive midrib and usually marginal hairs make this a very distinctive liverwort. The thallus, except for the midrib, is a single cell thick, another distinctive feature.

Similar Genera
Apometzgeria, another thallose genus with a stemlike midrib, is usually irregularly branched, and both surfaces are covered by hairs, while hairs in *Metzgeria* are confined to the underside of the midrib and the thallus margins.

Comments
The three species are usually easy to distinguish from each other: *M. temperata* has few marginal hairs, and the apex tapers to a point and bears numerous gemmae; *M. conjugata* has abundant short straight marginal hairs; and *M. leptoneura* has C-shaped marginal hairs.

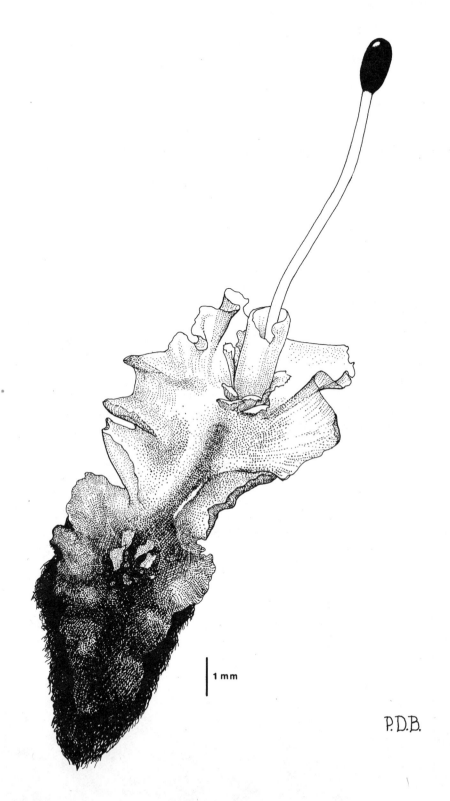

Moerckia blyttii from British Columbian specimen.

Moerckia Gott.

Name
In honor of A. Moerck, a collaborator in the production of the publication *Danish Flora*.

Species
Two species in the region.

Habit
Pale to dark green thalli, frequently with ruffled margins, firmly affixed to the substratum with colorless or reddish rhizoids.

Habitat
Usually on fine-textured, humus-rich substrata that are moist to wet most of the year, in open sites from sea level (especially near the coast) and alpine areas, but rare at mid-elevations.

Reproduction
Sporophytes occasional, maturing in summer to autumn; the elongate sporangia are dark brown to black, and open by one or two longitudinal lines.

Local Distribution
Probably widespread in oceanic climates and at high elevations in the region; most collections are from near the coast of Alaska and British Columbia, but the genus extends southward to Washington and Oregon.

World Distribution
The genus is widespread in the Northern Hemisphere in cooler climates in mountains, and moist climatic regions near sea level.

Distinguishing Characteristics
The upcurved ruffled margin of the thallus, the frequent presence of scales on the upper surface, and the elongate sporangium that usually opens by one or two longitudinal splits are all distinctive features. *Moerckia hibernica* thalli have a bandlike midrib, absent in most other local genera except *Calycularia*; in *C. crispula* there is a midrib (see below).

Similar Genera
Aneura sometimes has ruffled margins and grows in similar habitats, but the thalli tend to be rather more brittle and have thick rather than thin margins. The sporangium of *Aneura* opens by four lines. Thalli of *Pellia* can be of similar form, but lack any surface scales; the sporangia of *Pellia* are spherical. In the mountains, *Pellia* and *Moerckia* can strongly resemble each other when without sporophytes. Another important distinction other than sporangia of *Moerckia* is the antheridia, which in *Moerckia* are exposed and partially sheathed by a scale; in *Pellia* the antheridia are embedded within the thallus. *Calycularia* is similar in form, but has small scales on both the upper and lower surface; *Moerckia* has scales confined to the upper surface.

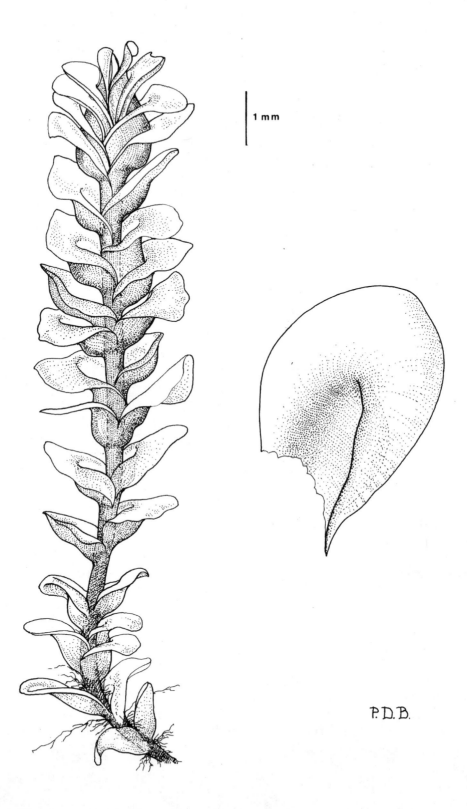

Mylia taylorii from British Columbian specimen.

Mylia S. Gray

Name
In honor of W. Mylius, a Dutch physician and patron of botany.

Species
Two species in the region.

Habit
Forms red brown (*M. taylorii*) to pale yellow green turfs (*M. anomala*) of erect to suberect plants, or with strands growing in *Sphagnum* tufts (*M. anomala*).

Habitat
In humid near-coastal areas characterized by high precipitation, both species are found in *Sphagnum* bogs, with *M. taylorii* more frequent in depressions and *M. anomala* in *Sphagnum* hummocks. In the forests, however, *M. taylorii* is often on logs and epiphytic on tree trunks. *M. taylorii* is also on damp cliff shelves, especially near waterfalls. Both species are mainly at low elevations but occur to subalpine elevations.

Reproduction
Both species produce sporophytes during the summer, but they are infrequently abundant. *M. anomala* usually produces abundant yellow green gemmae on the leaves and shoot apices. Sporangia are spherical.

Local Distribution
M. anomala is widespread in peatland throughout Alaska to northern California, while *M. taylorii* is predominantly near the coast of Alaska southward to Washington.

World Distribution
The genus is circumpolar in the Northern Hemisphere, most abundant near the coasts, especially in high-precipitation climates where there are mountains, but some species are also in bogs of more continental localities.

Distinguishing Characteristics
M. anomala is most frequently in *Sphagnum* hummocks in peat bogs, and generally produces numerous bright yellow green gemmae on the unlobed leaves. No other bog hepatic fits this description. *M. taylorii* is usually brilliant red orange, and this color, associated with the nearly round unlobed leaves, is usually enough to distinguish it.

Similar Genera
Jungermannia species also have unlobed leaves, but are usually half the size of *Mylia*, generally darker green, and never are red orange or produce numerous yellow green gemmae. *J. exsertifolia* is similar in size, but is dark green to nearly black and frequently grows on submerged rocks or wet cliffs associated with streams, especially at high elevations. *Nardia compressa* is also similar in size and form, but the color is frequently wine red, and the plants, like *J. exsertifolia*, grow commonly on seepy cliffs or in seepy areas at higher elevations. *Gyrothyra*, another genus with unlobed lateral leaves, sometimes takes on a reddish pigment. In these plants, however, the shoots tend to be reclining and strongly compressed (rather than erect and weakly compressed, as in *Mylia*), and the stems have rhizoids arising in tufts associated with purplish patches on the undersurface of the stem; such patches are absent in *Mylia*.

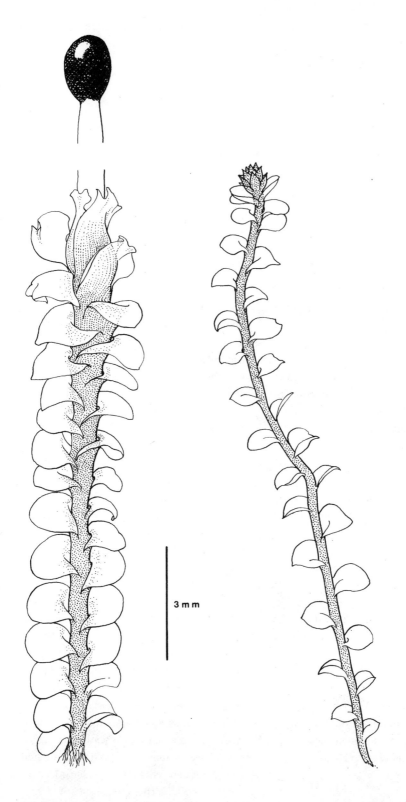

Nardia scalaris from British Columbian specimens.

Nardia S. Gray

Name
In honor of S. Nardi, an Italian patron of botany in the late seventeenth and early eighteenth centuries.

Species
At least six species in the region.

Habit
Diverse among the species, some creeping along the substratum and possessing slender dark green to reddish purple shoots, others forming tight, short turfs of pale green unbranched shoots, sometimes tinged with pink.

Habitat
Usually on mineral soil in areas where moisture is abundant. Thus in shallow pool depressions, late-snow areas, humid crevices of cliffs, especially near watercourses from near sea level to alpine elevations.

Reproduction
Most species produce sporophytes in spring or summer; those of pond depressions are often sterile, possibly reflecting the short season that the habitat is available after the pool water has disappeared.

Local Distribution
The genus is probably widespread in mountainous areas of the region and extends to near sea level along the coast from Alaska to northern California.

World Distribution
The genus is widespread in the Northern Hemisphere in arctic and temperate regions, frequent in mountains.

Distinguishing Characteristics
The small plants, the generally rounded leaves (even those with lobes), the incurved margins of the leaves, and the generally erect, unbranched or little-branched shoots are useful features.

Similar Genera
See notes under *Arnellia, Cryptocolea, Jungermannia, Mylia,* and *Odontoschisma*.

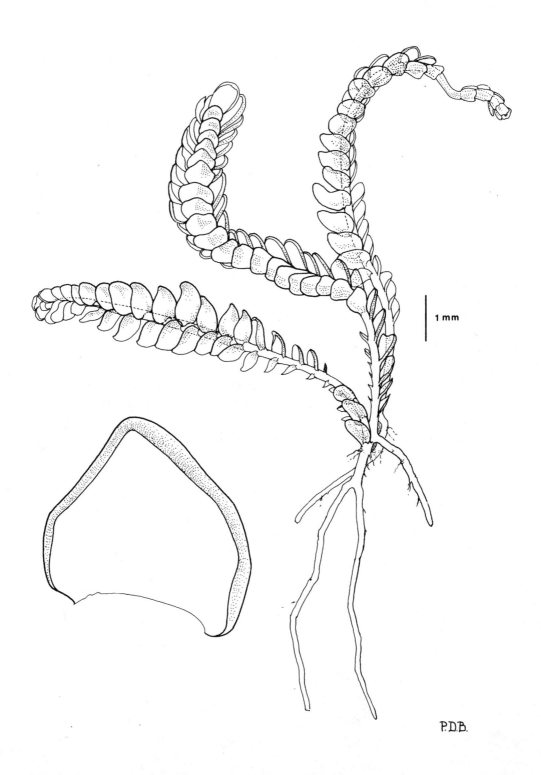

Odontoschisma denudatum from British Columbian specimen.

Odontoschisma (Dum.) Dum.

Name
Means "split-toothed," in reference to the toothed mouth of the perianth.

Species
Four species in the area.

Habit
Leafy shoots creeping to suberect, varying in size and color among the species, from orange yellow in *O. denudatum* to brown in *O. elongatum* and pale green to dark green in *O. sphagni* and *O. macounii*. Leaves decidedly concave, with margins incurved.

Habitat
Diverse; *O. macounii* on humus or mineral soil or on shaded rock faces, *O. sphagni* strands among *Sphagnum* shoots, *O. elongatum* on humic soil of peatland or humid outcrops; *O. denudatum* generally on tree trunks or on rotten logs, especially frequent on yellow cedar, from near sea level to subalpine elevations.

Reproduction
Gemmae are frequent and pale green in *O. denudatum* and *O. sphagni*. They have not been observed in local material of the other species. Sporophytes are not seen in local material.

Local Distribution
O. denudatum and *O. elongatum* are mainly near-coastal species from Alaska to Washington, while *O. sphagni* is both near the coast and in the interior of Alaska and northern British Columbia. *O. macounii* is a species of the north, in Alaska and British Columbia.

World Distribution
The genus is mainly tropical American with a few species found widely distributed in the Northern Hemisphere from arctic regions to warm temperate and tropical latitudes.

Distinguishing Characteristics
The small wormlike shoots with closely imbricated, concave, unlobed leaves make the genus relatively easy to separate. Many species have rootlike, leafless, rhizoid-covered shoots.

Similar Genera
Some *Jungermannia* species are similar in size, but leaves, although unlobed, are not strongly concave, and flagelliferous shoots are generally absent. See also *Cryptocolea*.

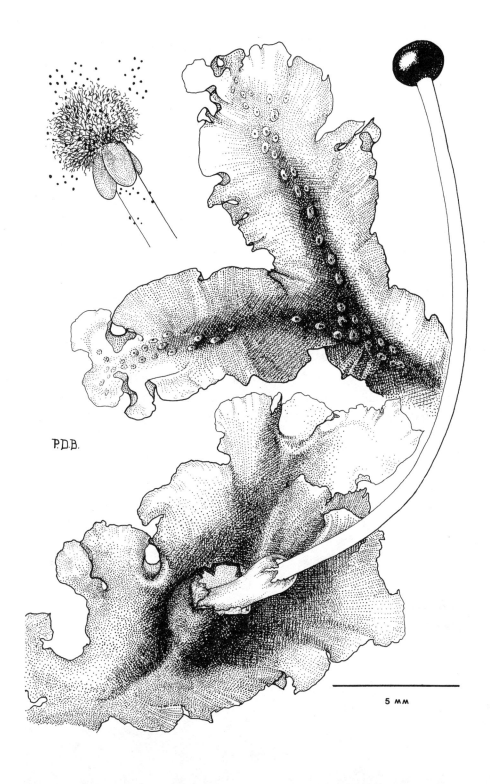

Pellia neesiana from British Columbian specimen. The male thallus is shown upper right.

Pellia Raddi

Name
In honor of L. Pelli-Fabbroni, a Florentine friend of the author of the genus.

Species
Three species in the region.

Habit
Dark green to pale green, flattened, usually somewhat smooth thalli, occasionally with purplish pigmentation (especially late or early in the season in some species); irregularly branched and forming extensive mats of overlapping lobes. Rhizoids sometimes purplish or brownish, especially near attachment to the thallus.

Habitat
Damp mineral soil of somewhat shaded moist to wet sites, especially near watercourses and ponds, from sea level to subalpine elevations.

Reproduction
Sporophytes abundant in spring to summer, the seta elongating rapidly and sometimes to 6 cm tall. When young, the seta is pale green, a feature infrequent in the liverworts. The opened sporangium, with its persistent furry mass of elaters, is striking.

Local Distribution
Widely distributed in the area, but infrequent in the drier climates.

World Distribution
The genus is widespread in temperate areas of the Northern Hemisphere.

Distinguishing Characteristics
The dark green thalli that sometimes have a purplish pigment in the midline, the flat rather than sinuous ruffled margins, the absence of any scales on the upper surface of the thallus, and the large seta and round sporangium that opens to expose a woolly mass of elaters that persist in the opened sporangium are all useful distinguishing features. Unfortunately, *Pellia*, especially in open sites in the mountains, has succulent thalli with ruffled margins, and thus resembles *Moerckia*. See notes under that genus.

Similar Genera
The absence of outlines of air chambers and their central pores immediately separate this genus from genera with thalli similar in size: *Reboulia, Conocephalum, Lunularia, Bucegia, Preissia,* and *Marchantia*. For thallose liverworts lacking such chambers and pores, and that are similar in form to *Pellia*, see notes under *Calycularia, Moerckia,* and *Aneura*.

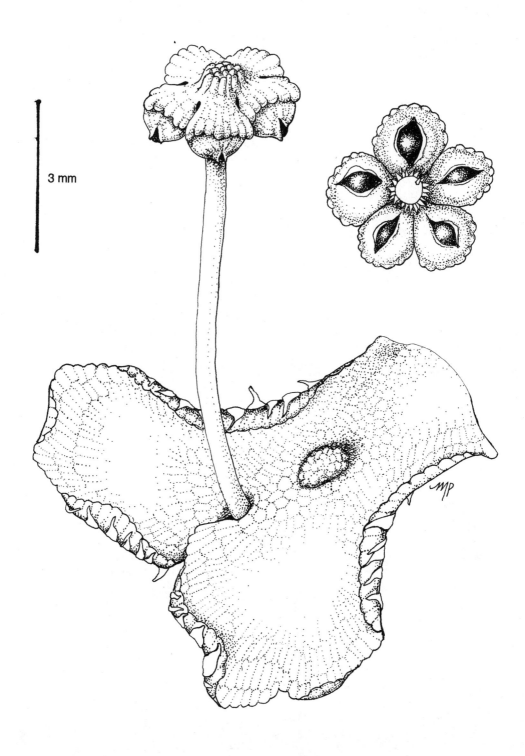

Peltolepis quadrata from British Columbian material amplified by illustrations of K. Mueller.

Peltolepis Lindb.

Name
Refers to the somewhat crescent-shaped scales on the undersurface of the thallus; literally "shield-scale."

Species
One species in the region: *P. quadrata*.

Habit
Dark green, small, flat thalli or with slightly upward arching margins from a depressed central, usually somewhat reticulate surface; the four- to six-lobed receptacles (five-lobed is usual) that bear the sporangia are raised 1 to 1.5 cm above the thallus; stalk of the receptacle purplish at base.

Habitat
On moist, open, fine-textured soil, especially in calcareous regions; also on cliff shelves and in crevices; in alpine areas.

Reproduction
Sporangia spherical and opening irregularly at the apex, mainly enclosed in the lobes of the receptacles.

Local Distribution
Rare in the northern and southern Rocky Mountains and in Alaska.

World Distribution
Rare, mainly in arctic and alpine regions in the Northern Hemisphere.

Distinguishing Characteristics
The scarcity of available specimens for study makes acquisition of field characteristics difficult. The presence of two rhizoid furrows on the stalk of the receptacle and the appendages (sometimes two) on the purplish ventral scales are useful distinguishing features.

Similar Genera
Sauteria and *Athalamia* are similar in form, but *Athalamia* lacks a furrow on the stalk of the receptacle, while *Sauteria* has a single furrow. In these genera the ventral scales lack appendages.

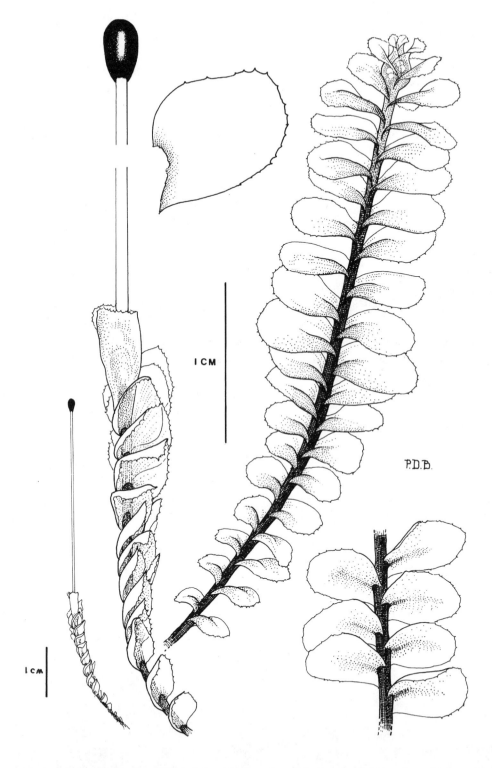

Plagiochila porelloides from British Columbian specimen. The upper-right specimen is viewed from above, while the specimen in the lower right is viewed from below.

Plagiochila (Dum.) Dum

Name
Describes the very flattened perianth.

Species
Five species in the region.

Habit
Leafy shoots usually unbranched or irregularly branched, varying from dark green to brownish green, suberect to reclining, the leaves somewhat recurved, especially when dry, stems varying from dark brown to green; the perianth is distinctly compressed and toothed at the mouth; in male plants the perigonial leaves are usually condensed in an apical or subapical elongate head and are pale green. Shoots form tall or low turfs.

Habitat
Highly variable: epiphytic up tree bases, over soil on rocks and directly on soil, especially near watercourses; on humid cliffs and sometimes in swampy areas, from sea level to subalpine and alpine elevations.

Reproduction
Sporophytes maturing in spring to summer; not common.

Local Distribution
The genus widely distributed in the region from sea level to alpine elevations. Only *P. porelloides* is frequent; *P. arctica* and *P. schofieldiana* are confined to Alaska and British Columbia, while *P. semidecurrens* is mainly near the coast from Alaska to northern Oregon. *P. poeltii* is in several of the Aleutian Islands.

World Distribution
The genus shows a wide distribution throughout the world.

Distinguishing Characteristics
The asymmetric unlobed leaf, the strongly decurrent leaf base on the upper surface of the shoot, and the slight recurving of the leaf from a longitudinal groove up from the leaf base make this liverwort very distinctive. All of our species have some teeth on the leaf margin, although in *P. arctica*, *P. poeltii*, and some specimens of *P. porelloides* these teeth can be very blunt. When the perianth is present, the compressed mouth is very distinctive.

Similar Genera
Other genera with asymmetric lateral leaves similar to *Plagiochila* possess underleaves, but these may be obscure. In these genera, however, including *Lophocolea* and *Geocalyx*, the color tends to be pale to yellow green, and perianths are not laterally compressed at the mouth; these genera, too, tend to have leaves with a sinus between the apical lobes. *Harpanthus*, too, has underleaves and a notched apex of the lateral leaf; this hepatic tends to be darker green. *Chiloscyphus*, although darker or sometimes much paler green, tends to have the leaves essentially entire and is usually in a very wet or aquatic environment; most species of *Plagiochila* are not in extremely wet environments.

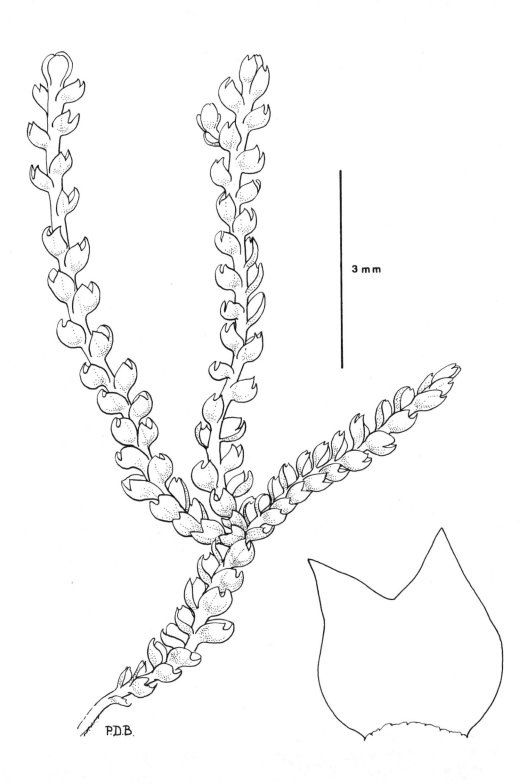

Pleurocladula albescens from Washington State specimen.

Pleurocladula Grolle

Name
Describes the way the lateral branches emerge, compared to the genus *Cephalozia*, to which it is related.

Species
One species in the region: *P. albescens*.

Habit
Forming dense, very pale green turfs of interwoven, irregularly branched shoots in which the bilobed leaves diverge outward, pouchlike, from the stem; frequently intermixed among other bryophytes or herbs.

Habitat
Subalpine to alpine late-snow areas, near ponds, and on damp cliff ledges, usually where persistent moisture is available.

Reproduction
Neither sporophytes nor gemmae have been noted in local material; shoots are very brittle and probably serve in propagation.

Local Distribution
Widely distributed at high elevations from Alaska to California.

World Distribution
Circumpolar in arctic and mountainous regions of the Northern Hemisphere.

Distinguishing Characteristics
The strongly incurved, saclike, bilobed pale green leaves make the shoots appear somewhat wormlike. This, associated with the high-elevation, late-snow-bed habitat, is usually enough to distinguish this liverwort.

Similar Genera
Cephalozia is similar in size, but the leaves are not remarkably saclike or the shoots wormlike; usually the plants are not whitish green, as in *Pleurocladula*. Although *Gymnomitrion* shoots are often wormlike, the plants show little evidence of green; the same is true for *Anthelia*. *Gymnomitrion* usually occurs on dry rock surfaces, while *Pleurocladula* is terrestrial in wet sites. In *Anthelia* and *Gymnomitrion* the leaves are closely overlapping, while in *Pleurocladula* the leaves barely overlap.

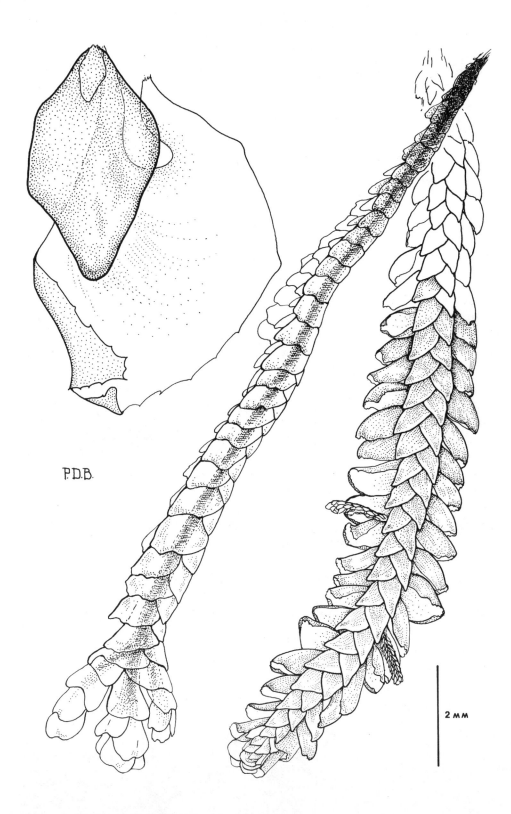

Pleurozia purpurea from British Columbian specimen.

Pleurozia Dum

Name
Means "side bud," in reference to the lateral budlike perianths.

Species
One species in the region: *P. purpurea*.

Habit
Forming bright reddish to orange purple turfs or mats of erect to reclining unbranched shoots; usually in pure colonies, but sometimes intermixed with other bryophytes.

Habitat
Usually on cliff ledges in open sites near sea level or on slopes or in depressions of peatland; sometimes on logs and occasionally at tree bases, especially in humid, open coniferous forest, confined to extremely humid climates.

Reproduction
Sporophytes and gemmae unknown in local material.

Local Distribution
Confined to low elevations of the open coast from northern Vancouver Island, British Columbia, to southeastern Alaska, extending sporadically along the Aleutian Chain.

World Distribution
Pleurozia purpurea shows a very interrupted distribution through the Northern Hemisphere, being confined to strongly oceanic climates or areas of high precipitation: northwestern coast and islands of Europe, the Himalayas, Japan, and the northwest coast of North America. The genus is mainly one of tropical latitudes.

Distinguishing Characteristics
The bright wine red to red orange plants that bear markedly inflated lobes and lobules of the leaves make this plant unmistakable.

Similar Genera
Although some species of *Scapania* are wine-red, the lobes and lobules are never inflated.

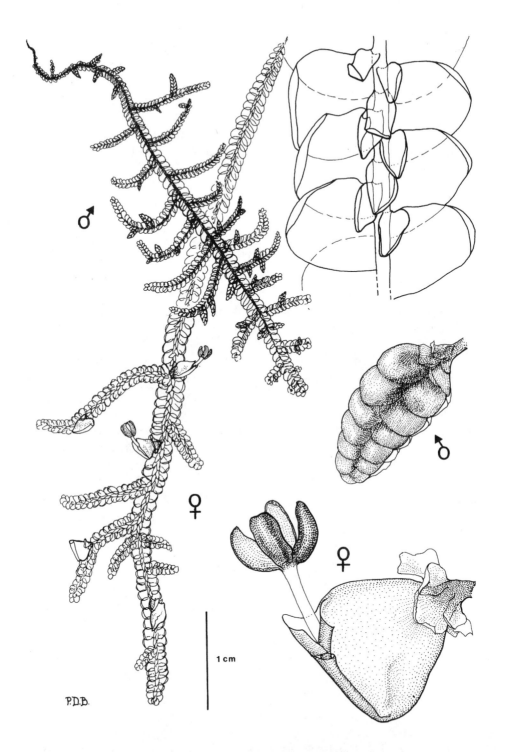

Porella navicularis from British Columbian specimen. The upper right shows the shoot, apex downward, viewed toward the undersurface, showing lobes and lobules of lateral leaves, and the unlobed underleaves. The female plant bears several perianths with emerging sporophytes, while the male plant has many small catkin-like branches that bear the sex organs.

Porella L.

Name
Means "small opening," perhaps in reference to the pinched, flattened mouth of the perianth.

Species
Six species in the region.

Habit
Dark to light green to rusty brown, glossy or dull, usually regularly pinnate plants, forming compressed masses in which the shoots hang with the apex downward and sometimes arching outward. The flattened perianths are on the undersides of the shoots, while the perigonia (antheridial branches) form catkin-like branchlets on the lateral branches and are pale green.

Habitat
On perpendicular shaded surfaces of boulders and outcrops; also commonly epiphytic, especially on broadleafed maple, oak, red alder, and California Bay.

Reproduction
Sporophytes abundant in winter to spring, especially in near-coastal populations.

Local Distribution
The genus is widely distributed from the Aleutian Chain of Alaska southward to California.

World Distribution
The genus is mainly tropical, but several species are confined to temperate areas. Most of our local species are confined to western North America.

Distinguishing Characteristics
The usual pinnate branching of plants that curve outward with a convex curve from the substratum, the unequally bilobed leaf in which the lobule is on the ventral surface and similar in form to the underleaf, and the somewhat compressed mouth of the perianth are very distinctive features.

Similar Genera
Frullania is smaller, the underleaves are equally bilobed (unlobed in *Porella*), and the lobule is helmet-shaped (recurved margins in *Porella*). *Radula* is usually smaller and lacks underleaves. *Ascidiota* tends to be nearly black, and the leaf margins are ciliate-toothed, a feature less developed in a single local species of *Porella*: *P. vernicosa*.

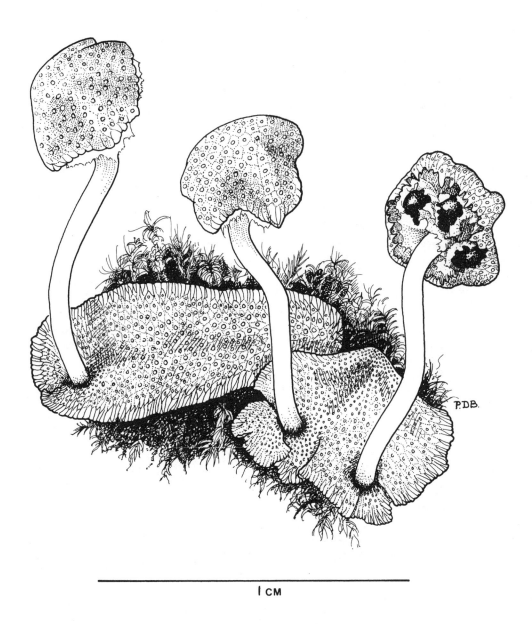

Preissia quadrata from British Columbian specimen.

Preissia Corda

Name
In honor of B. Preiss, a nineteenth-century Prague physician.

Species
One species in the region: *P. quadrata*.

Habit
Pale green thalli firmly affixed to the substratum; usually forming mats of interwoven and overlapping lobes; the pores are usually apparent under the hand lens; under the hand lens the pore shows a crosslike opening formed by the four swollen cells that bound the inner part of the opening. Scales and undersurface are reddish brown.

Habitat
On mineral soil, most frequently in calcium-rich sites, especially near watercourses, predominantly at lower elevations, but extending to subalpine and alpine elevations.

Reproduction
Sporophytes frequent in spring; the sporangium-bearing receptacle is often very regularly four-lobed and somewhat angled, and the pores are usually apparent on its pale green surface. The antheridial receptacle is usually dark purple and has a somewhat undulate margin.

Local Distribution
Widely distributed in the region, especially in calcium-rich bedrock areas, from sea level to subalpine.

World Distribution
Arctic and temperate areas of the Northern Hemisphere.

Distinguishing Characteristics
The usually four-lobed angular sporophyte-bearing receptacles, the pale green color of the plants, the pores that appear to have a crosslike opening within the pore, the usually lime-rich substratum, and the purplish scales are usually enough to separate this genus.

Similar Genera
See notes under *Bucegia*.

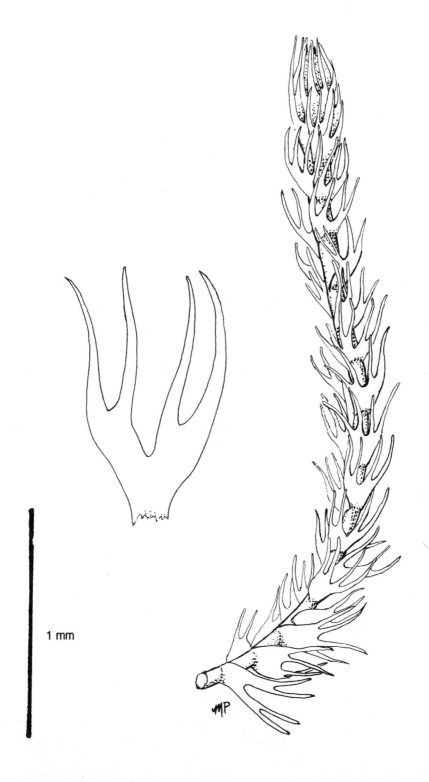

Pseudolepicolea fryei from Alaskan specimens enhanced by reference to illustrations by R. M. Schuster.

Pseudolepicolea Fulf. & J. Tayl.

Name
Means "resembling *Lepicolea*"; *Lepicolea* is a genus of the Southern Hemisphere.

Species
A single species in the area: *P. fryei*.

Habit
Slender pale brownish to green erect or sub-erect, sparingly and irregularly branched plants, occurring as individual strands or in turfs.

Habitat
Very wet acidic tundra, often with *Sphagnum*.

Reproduction
Sporangium yellow brown, broadly ellipsoidal, emerging from an elongate, deeply four-pleated perianth.

Local Distribution
Mainly in the Bering Sea area of northwestern Alaska, extending from coastal islands into the Brooks Range.

World Distribution
The local species also extending to the west coast of Hudson Bay; the genus found also in southern South America and in Southeast Asia.

Distinguishing Characteristics
The deeply bilobed leaves in which each of the lobes is also bilobed make this a very distinctive plant.

Similar Genera
Blepharostoma arachnoideum has leaves in which the lobes are forked, but in this species the leaf lobes are a single cell in width, while in *Pseudolepicolea* the lobes are three to four cells in width. *Herbertus aduncus* has some environmental variants that are extremely thread-like, but the leaves are bilobed with no forking of the lobes; plants are also dark brown to reddish brown rather than somewhat brownish to green, as in *Pseudolepicolea*. *Tetralophozia* has teeth on the consistently equally four-lobed leaves, rather than two-lobed with each fork, as in *Pseudolepicolea*. *Kurzia* and *Lepidozia*, although sometimes similar in size to *Pseudolepicolea*, have three- to four-lobed leaves rather than forked, two-lobed leaves, as in *Pseudolepicolea*, and these have side branches arising at right angles to the main shoot, a feature absent in *Pseudolepicolea*.

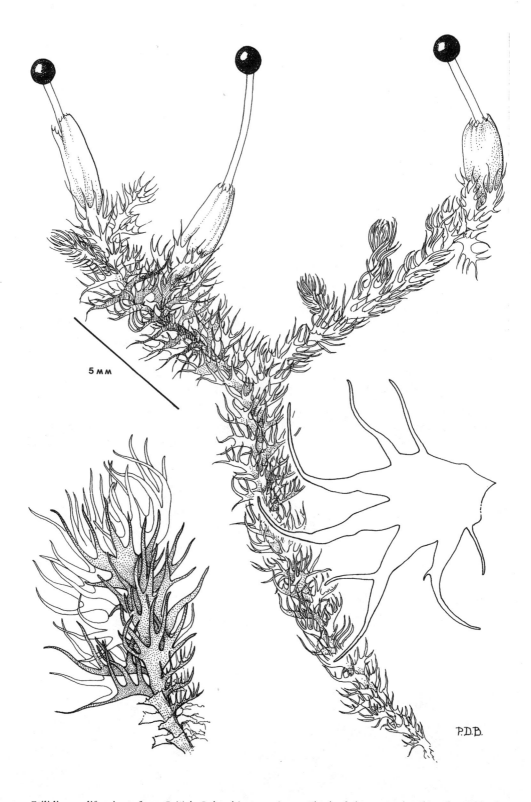

Ptilidium californicum from British Columbian specimen. The leaf shown in detail is a lateral leaf.

Ptilidium Nees

Name
Means "small feather," in reference to the feathery segments of the leaves and possibly the feathery appearance of many plants.

Species
Three species in the region.

Habit
Rusty brown, reddish brown, orange brown to dark green, in two species firmly affixed to the substratum, but in *P. ciliare* sometimes forming loose mats or turfs of erect to suberect plants. Branching regular to irregular.

Habitat
Epiphytic on living trees, often on rotting logs or on boulders or cliff faces, in *P. ciliare* among boulders of boulder slopes or among shrubs, from sea level to tundra in alpine elevations; occasionally in bogs where it can be aquatic for part of the growing season.

Reproduction
P. pulcherrimum and *P. californicum* produce abundant sporophytes from spring to autumn, dependent on the altitude of their location. *P. ciliare* rarely produces sporophytes in the region.

Local Distribution
P. ciliare is scattered throughout Alaska and British Columbia, while *P. pulcherrimum* seems to be confined to southern coastal Alaska southward to Washington. *P. californicum* is predominantly coastal and in subalpine forest both near the coast and in interior mountains from Alaska to northern California.

World Distribution
Widespread in the Northern Hemisphere, extending southward from arctic to temperate regions and reappearing erratically in the Southern Hemisphere.

Distinguishing Characteristics
The many-lobed leaves with ciliate or finger-like teeth, the soft, usually brownish red to orange plants are usually enough to separate this genus.

Similar Genera
See notes under *Mastigophora* and *Ascidiota*.

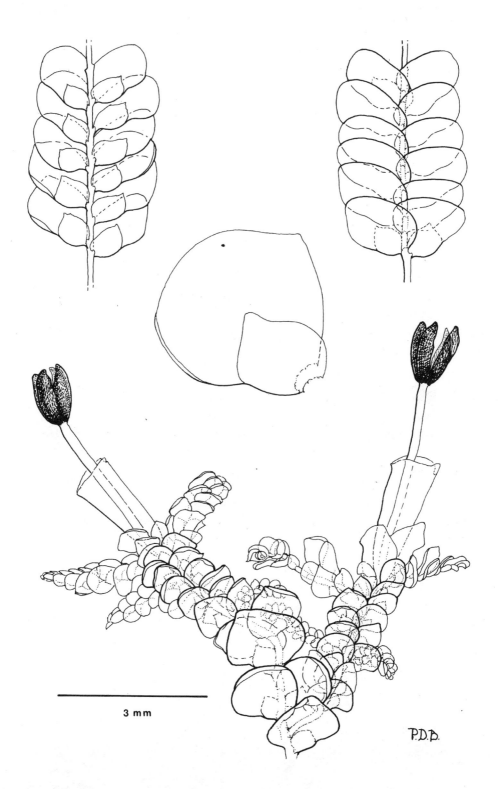

Radula complanata from British Columbian specimen. Ventral view on upper left; dorsal view on upper right.

Radula Dum.

Name
Means "scraper," in reference to the flattened perianth mouth that resembles a scraper.

Species
Six species in the region.

Habit
Usually glossy yellow green but sometimes pale brownish to dark brownish, regularly pinnate to irregularly branched, usually firmly affixed to the substratum, but sometimes loose.

Habitat
Epiphytic on living trees, also on rock surfaces, cliff ledges, and terrestrial among other plants, especially in humid environments; from near sea level to alpine elevations.

Reproduction
Sporophytes frequent in spring and persisting for the entire year in some species; gemmae are sometimes abundant on vegetative plants, where they form dusty bright yellow green masses on the leaf margins. Sporophytes and gemmae not noted in some local species. The plants are often brittle; fragments can serve to disseminate the species.

Local Distribution
Widely distributed in the region, but frequent only near the coast, with *R. complanata* as the most widespread species.

World Distribution
The genus is widely distributed throughout the world.

Distinguishing Characteristics
The small, bulging lobule and the absence of underleaves are usually sufficient to separate this genus. Most species are pale to yellow green and glossy.

Similar Genera
Cololejeunea is extremely similar in size and appearance to *Radula bolanderi*. The plants, however, tend to be bright yellow green and are never shiny when dry, whereas *Radula* is somewhat shiny. The lobule of *Radula* is also more pronounced than that of *Cololejeunea*. *Radula brunnea* and *R. auriculata* superficially resemble a large *Frullania*, but lack the underleaves and helmet-shaped lobules of *Frullania*. *R. auriculata* resembles a small *Porella*, but lacks the large lobules and underleaves of *Porella*.

Comments
The presence of *Radula brunnea* on Saddle Mountain, Oregon, suggests a remnant of an extremely ancient flora of East Asia relationships; the species is otherwise restricted to Southeast Asia, and its nearest relatives are confined mainly to the tropics.

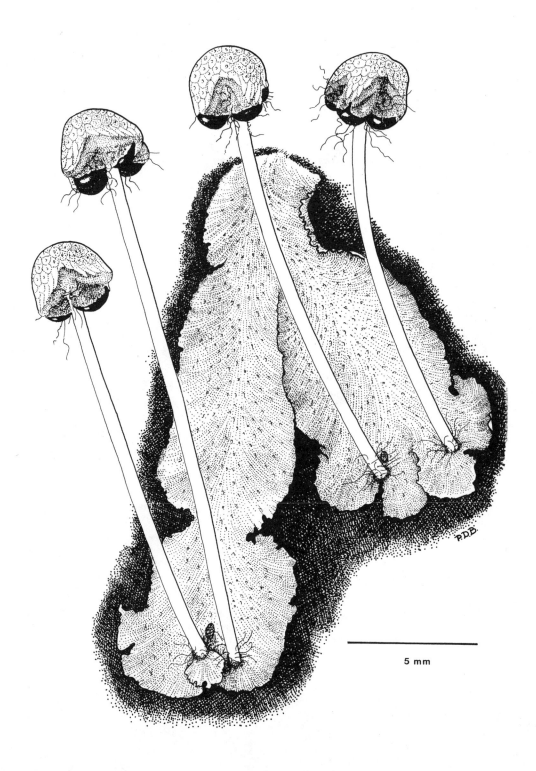

Reboulia hemisphaerica from British Columbian specimen.

Reboulia Raddi

Name
In honor of E. de Reboul, a nineteenth-century Florentine botanist.

Species
One species in the region: *R. hemisphaerica*.

Habit
Flattened, light green thalli with purple margins and dark purple scales and undersurfaces; the pores and air-chamber pattern are apparent under the hand lens.

Habitat
On earth of somewhat shaded ledges of cliffs, predominantly at lower elevations.

Reproduction
Sporophytes maturing in spring; the subspherical sporangia open by a lid rather than by longitudinal lines.

Local Distribution
Probably widespread from southeastern Alaska to California.

World Distribution
Widely distributed in temperate and tropical regions.

Distinguishing Characteristics
The purple margins, deep purple scales, the relatively flat thallus when dry, and the somewhat rounded sporangium-bearing receptacle with a stout stalk and sporangia that open by an irregular lid are all useful distinguishing characters.

Similar Genera
Other thallose genera that have dorsal pores and are similar in size are *Targionia, Mannia, Asterella, Preissia,* and *Bucegia*. In *Targionia* and *Mannia* the thalli are strongly incurved when dry; the receptacles of *Mannia* have a very slender stalk, while the sporangia of *Targionia* are ventral and are not on a receptacle. *Asterella* also has a slender stalk to the receptacle, and each sporangium is fringed by a white involucre, lacking in *Reboulia*. *Preissia* and *Bucegia* show clear barrel pores with four swollen cells in the inner part of the pore; in *Reboulia* the pores are simple, thus lack these bulging cells. The receptacle of *Reboulia* and the caplike opening of the sporangium differ from the longitudinal opening of the sporangium in the other two genera.

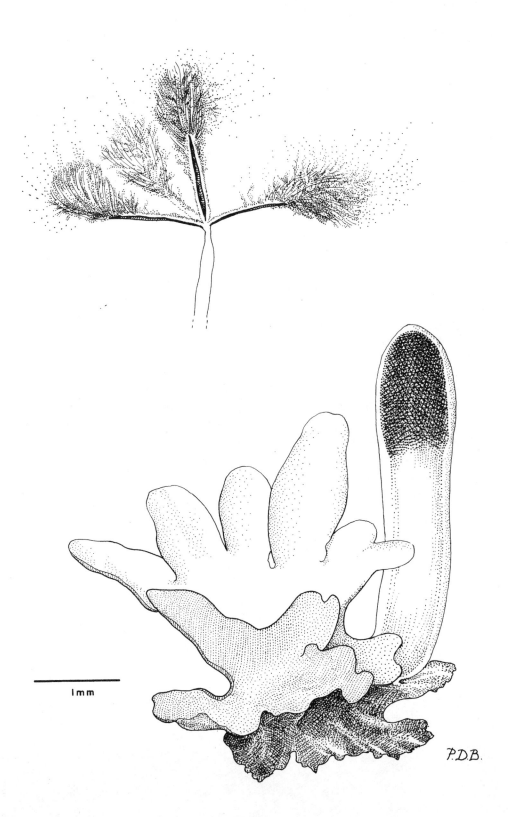

Riccardia latifrons from British Columbian specimen.

Riccardia S. Gray

Name
In honor of V. Riccardi, a nineteenth-century Florentine patron of botany.

Species
Four species in the region.

Habit
Dark green, regularly to irregularly branched thalli forming short turfs or mats.

Habitat
Rotten logs, moist rocks, damp soil, in somewhat open areas within forests or in non-forested areas, also on cliff shelves, especially in humid to waterfall-spray sites, and in swamps and peatland.

Reproduction
Sporophytes frequent in early spring to summer. The opened sporangia are very striking with tufts of elaters at the tips of the four divisions of the sporangium. The sleeve that encloses the developing sporangium is pale green, and the sporophyte is visible within this sleeve before the seta elongates to push the sporangium well above the thallus surface. Gemmae are sometimes abundant, emerging from the tips of lobes, but these are not obvious, even with a hand lens.

Local Distribution
The genus is widely distributed in the region, particularly in humid forests, and absent in drier climatic regions.

World Distribution
The genus is widely distributed in the world; it is especially rich in species in the tropics and subtropics.

Distinguishing Characteristics
The usually dark green succulent thalli with very narrow lobes, the rather small size (lobes usually 1 to 3 mm in diameter, and thalli are usually 1 to 2 cm long), and the thallus of nearly the same thickness through its width are useful characteristics.

Similar Genera
Aneura thalli tend to be larger (usually 4 to 5 mm in diameter) and to have few lobes, and the thallus tends to be texturally slippery with its edges curving upward somewhat; it thus differs from all local species of *Riccardia*. *Riccia fluitans* can resemble *Riccardia*, but a cross section of the thallus will show the *Riccia* to have internal air chambers, and the thallus, when dry, is papery and dark green to light green; in *Riccardia* dry thalli are brittle and nearly black.

Riccia sorocarpa from British Columbian specimen.

Riccia L.

Name
In honor of P. F. Ricci, an eighteenth-century amateur botanist of Florence.

Species
Fifteen species in the region.

Habit
Thalli small, usually very regularly forked, with the branches generally 2 to 4 mm wide or less, varying from pale green to bluish green or dark green (*R. fluitans*).

Habitat
R. fluitans is a submerged aquatic, usually floating under the water surface in quiet pools, while all other species are terrestrial on mineral soil of open sites in drier milder climates and usually are visible only during the spring; raw mineral soil of exposed sites on outcrop terraces and field margins are common habitats, but in the southern part of the range they are on earth of semiarid areas where moisture is available in winter or spring.

Reproduction
Sporophytes common in spring, within the thallus and not exposed until the thallus decomposes. Fragmentation is undoubtedly important in *Riccia fluitans*.

Local Distribution
Predominantly near the coast from southwestern British Columbia to southern California, reaching its highest diversity in California, but one species rare in subalpine northern Alaska.

World Distribution
The genus is widely distributed in drier climatic regions of the world from tropical to cool temperature latitudes, but some species are found also in more humid climates.

Distinguishing Characteristics
In most species of *Riccia* the thalli are small and forked, often growing in rosettes of roughly heart-shaped thallus tips on soil, and have a non-glossy, somewhat bluish green appearance. In *R. fluitans*, however, the thallus resembles *Riccardia* in some respects. The tiny thalli with internal sporangia are very distinctive.

Similar Genera
See notes under *Riccardia*. *Ricciocarpos* thalli are much larger than *Riccia*; in aquatic forms of *Ricciocarpos* the ventral scales are very conspicuous. Such are lacking in *Riccia*.

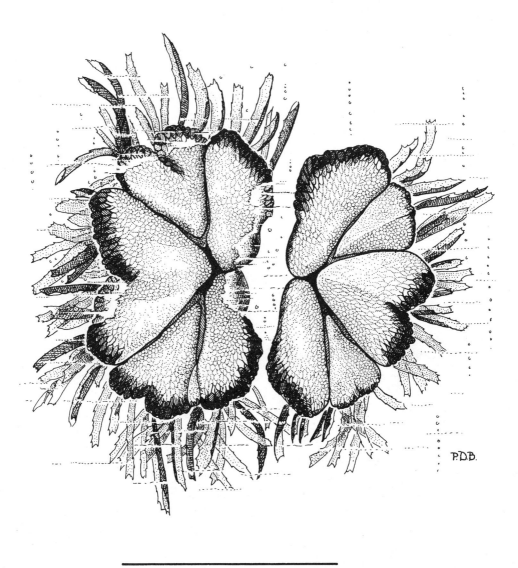

Ricciocarpos natans from British Columbian specimen.

Ricciocarpos Corda

Name
Refers to the resemblance of the genus to *Riccia*.

Species
A single species in the region: *R. natans*.

Habit
Usually floating on the surface of quiet water, especially in areas where organic pollutants are present; also growing on mud when stranded by withdrawal of the water from the aquatic site. The thalli tend to be regularly heart-shaped and bright green with large masses of purple scales beneath, which serve to give the floating thallus stability.

Habitat
Backwaters, ponds, and coves of lakes and watercourses.

Reproduction
Sporophytes infrequent. Presumably propagated mainly by overwintering fragments of thallus lobes.

Local Distribution
Widely scattered through the area and abundant locally, but a few localities have been documented by specimens from Alaska to California.

World Distribution
Widely distributed throughout the world, except in polar climates.

Distinguishing Characteristics
No other genus of local liverworts forms floating half-rosettes. The forked heart-shaped lobes with purplish midline and long banner-like stabilizing scales on the undersurface are very distinctive features.

Similar Genera
No other genus in the region forms surface-floating thalli. *Riccia cavernosa* that grows sometimes on soil at the margins of water bodies is similar in size to *Ricciocarpos* thalli that also can grow on these sites. In *Ricciocarpos*, however, the thallus surface tends to be darkish green, and retains the purplish midline; in the *Riccia* the thallus is very pale green and appears somewhat spongy under the hand lens.

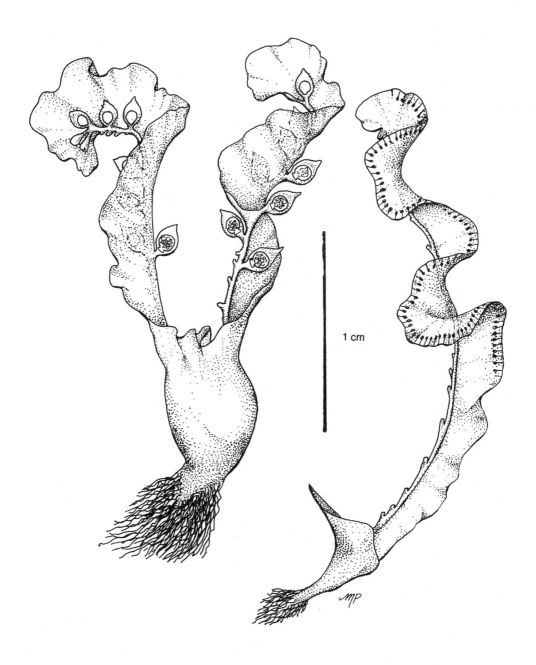

Riella affinis from a Texan specimen. The left figure has sporophytes; the right figure is male.

Riella Mont.

Name
In honor of Du Rieu de Maisonneuve, a nineteenth-century director of the Botanical Garden in Bordeaux, France.

Species
Two species in the area.

Habit
Thalli 6 to 30 mm high, forming dense masses, unbranched or occasionally forked, erect flattened stem with frill-like wing 1 to 3 mm wide, green and translucent, attached to mud by rhizoids from the lower portion of the stem. Resembling a thallose green alga.

Habitat
Submerged, attached to mud of ephemeral water bodies, apparent as the water level drops, possibly confined to the shallower marginal areas.

Reproduction
Sporophytes abundant in marginal involucres opposite the wing of the thallus, maturing in spring; gemmae also present on the stem.

Local Distribution
The two local species noted only from California, where rare and local.

World Distribution
The genus is in North America; also in Texas and South Dakota and outside North America in the Mediterranean area, in South America, the Canary Islands, South Africa, and Australia.

Distinguishing Characteristics
The submerged aquatic habitat combined with the one-winged thallus with sporophytes in pleated involucres on the stem make this highly distinctive.

Similar Genera
No local genus is likely to be confused with *Riella*.

Comments
This liverwort resembles a green alga in its paper-thin erect thalli attached to the bottom mud of water bodies of fluctuating water level.

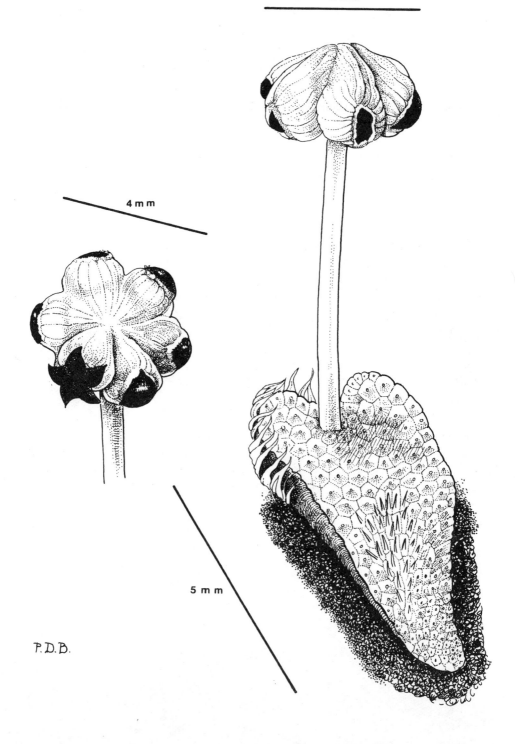

Sauteria alpina from Alaskan specimen amplified with illustrations of D. Shimizu.

Sauteria Nees

Name
In honor of A. Sauter, a nineteenth-century Austrian physician.

Species
One species in the region: *S. alpina*.

Habit
Pale green to bluish green small thalli with air pores and chambers apparent with hand-lens magnification, bearing green receptacles that are regularly five- to six-lobed, with colorless ventral scales.

Habitat
On mineral soil, generally calcium-rich, in alpine or high-latitude localities, generally near cliff bases or on cliff ledges.

Reproduction
Receptacles bearing spherical sporangia that open with four to five longitudinal lines, maturing in summer.

Local Distribution
Extremely rarely collected in mountains of Alaska and British Columbia.

World Distribution
S. alpina widely distributed in high mountains and arctic areas of the Northern Hemisphere; the other species, *S. berteroana*, is in Chile.

Distinguishing Characteristics
Based on hand-lens characteristics, this genus is difficult to distinguish from *Athalamia* and *Peltolepis*. In *Athalamia*, however, the spores are rusty, while in the other genera spores are dark to blackish brown.

Similar Genera
See note above.

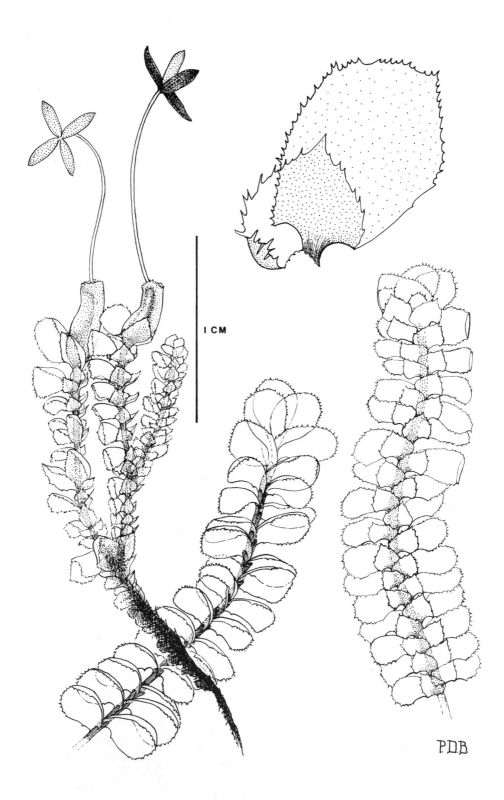

Scapania bolanderi from British Columbian specimen.

Scapania (Dum.) Dum.

Name
Describes the spadelike flattened perianth.

Species
Thirty-six species in the region.

Habit
Most species grow in turfs or tufts of reclining to suberect unbranched or irregularly branched plants. Color varies from pale green through dark green to reddish purple, dark brown, or nearly black. Size is equally variable, with some species having leafy shoots less than 2 mm in diameter and less than 1 cm long, while others have shoots to nearly 1 cm in diameter and more than 10 cm. long. A few species have the lobule nearly as large as the lobe, and thus pose problems in generic recognition.

Habitat
The many species show a considerable diversity of habitat; some are epiphytic or on rotten logs in forests, others are mainly on wet cliffs; some are on wet rocks or soil near or in lakes and streams; still others are confined to wet or boggy sites. They are found from near sea level to alpine elevations.

Reproduction
Sporophyte production occurs abundantly in some species in spring, while others produce them in summer. Some species have never been noted to produce sporophytes. Some species produce gemmae on the leaf margins.

Local Distribution
The genus is widely distributed in the region, reaching its greatest diversity in Alaska.

World Distribution
The genus is predominantly in the Northern Hemisphere, but a few species are in the Southern Hemisphere.

Distinguishing Characteristics
The distinctive overlapping, generally somewhat rounded lobule on the dorsal surface is usually sufficient to distinguish the genus. In most species the larger lobe is slightly recurved.

Similar Genera
Some specimens of *Diplophyllum* resemble *Scapania*, but the lobule tends to be elongate and not circular in outline; this gives specimens of *Diplophyllum* a more compressed appearance. *Douinia*, although it has a lobule similar to that of *Scapania*, usually grows in far drier habitats, and the plants tend to be smaller than most specimens of *Scapania*. In *Douinia* the perianths are not compressed, the usual condition in *Scapania*. In *Pleurozia* both lobe and lobule are inflated, never so in *Scapania*. Some specimens of *Lophozia* have a somewhat smaller lobe on the dorsal surface, and the plants are not dorsiventrally compressed, the usual condition in *Scapania*.

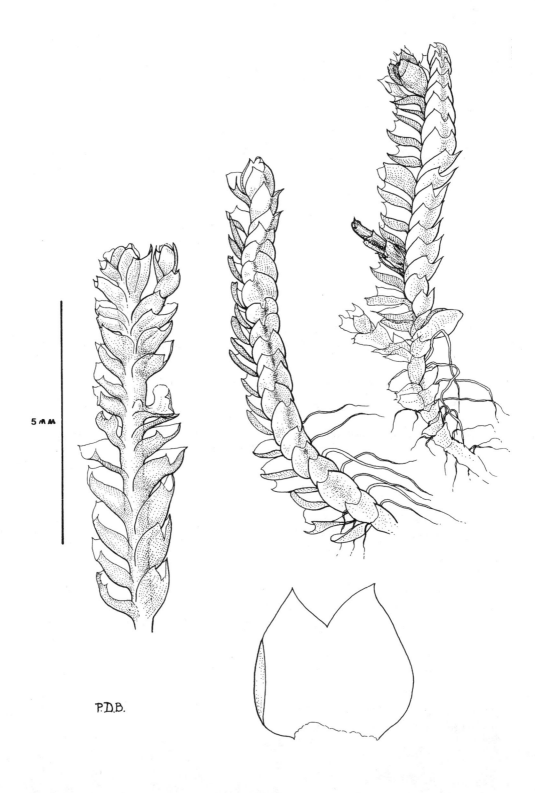

Schofieldia monticola from Washington State specimen.

Schofieldia J. Godfr.

Name
In honor of W. B. Schofield, a twentieth-century Canadian botanist who first collected the plant.

Species
One species in the region: *S. monticola*.

Habit
Forming dense turfs or mats of suberect or reclining shoots, pale to dark green and succulent, sometimes appearing somewhat compressed.

Habitat
Seepage slopes of late-snow areas in subalpine to alpine regions, often in somewhat shaded sites among heath shrubs on slopes.

Reproduction
Sporophytes apparently rare, maturing in autumn. Gemmae are produced by leaves near shoot apices.

Local and World Distributions
Endemic to western North America. Confined to subalpine and alpine elevations, mainly near the coast, from southeastern Alaska southward to Oregon.

Distinguishing Characteristics
The somewhat succulent dark green to pale green plants that form tight turfs combined with the high elevation sites and late-snow-bed habitat are useful features.

Similar Genera
The genus resembles some species of *Lophozia*, even the somewhat succulent nature of the plants. In *Lophozia*, however, the plants are not watery dark green but are generally pale green or yellowish green; those that are dark green tend to have thin, not succulent, leaves.

Sphaerocarpos texanus from British Columbian specimen.
A male plant is surrounded by female plants.

Sphaerocarpos Boehmer

Name
Describes the spherical sporangium, often enclosed in a subspherical sheath.

Species
Five species in the region.

Habit
Tiny pale green thalli with ruffled margins; the inflated female involucres obscuring the thallus; male involucres often purplish.

Habitat
Fine-textured soil of terraces on outcrops and on shallow soil over horizontal outcrops, especially in areas where moisture persists longer in the spring; also in the Californian portion of its range a plant of ditches, trail margins, and gardens.

Reproduction
Sporophytes common, when mature giving the protective sleeve a darkened color.

Local Distribution
Rare in southwestern British Columbia and adjacent Washington, more frequent southward to California, where it reaches its highest diversity.

World Distribution
Mainly in warm temperate climates in both Northern and Southern Hemispheres.

Distinguishing Characteristics
The small pale green thalli with saclike involucres that contain a single spherical sporophyte are very distinctive.

Similar Genera
Only *Geothallus* is likely to be confused with *Sphaerocarpos* (see notes under that genus).

Comments
This genus is one of the few bryophytes on which genetic studies have been made. Spore ornamentation is inherited through the female parent. Two of every four spores grow into male plants, while the other two form females.

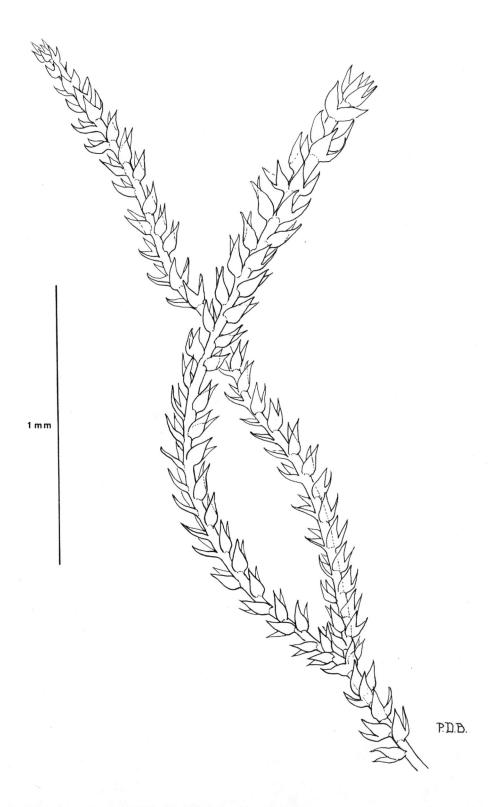

Sphenolobopsis pearsonii from British Columbian specimen.

Sphenolobopsis Schust. & Kitag.

Name
Noting its resemblance to *Sphenolobus*, the name is derived from the leaf that appears V-shaped when viewed from above; literally "wedge-flap."

Species
One species in the region: *S. pearsonii*.

Habit
Forms thin reddish brown to dull yellowish green filaments or patches over rock surfaces.

Habitat
Shaded outcrop or boulder faces from near sea level to subalpine areas near the coast.

Reproduction
Sporophytes not seen in local material.

Local Distribution
Very rare; a few collections from near the coast in the highest precipitation regions in southeastern Alaska and adjacent British Columbia.

World Distribution
Showing a very interrupted distribution in high moisture climates in the Northern Hemisphere; considered rare throughout its range.

Distinguishing Characteristics
The small size, the reddish brown color, and the shaded, well-drained, rock-surface habitat are usually sufficient to determine this genus.

Similar Genera
Several genera have species similar in size and appearance to *Sphenolobopsis*; these, however, are rarely present on shaded rock surfaces that are relatively dry. See *Cephaloziella*, *Marsupella*, and *Eremonotus*.

Takakia lepidozioides from British Columbian specimen.

Takakia Hatt. & H. Inoue

Name
In honor of the twentieth-century Japanese botanist N. Takaki, who made the collections on which the genus was recognized.

Species
Two species in the region.

Habit
Forming bright green turfs of erect plants usually less than 2 cm tall; colorless to pale green reclining leafless shoots often abundant; usually loosely affixed to substratum by a creeping intertangled rootlike system.

Habitat
Humus or peaty banks, usually on sloping or perpendicular surfaces that are somewhat shaded and sheltered, often in sites where the air is moist, as near streams, pools, cascades, and waterfalls, from near sea level to alpine elevations.

Reproduction
All parts of the plant are brittle and presumably serve in vegetative propagation; sporophytes of *T. ceratophylla* have a persistent seta, the elongate sporangium bears an apical calyptra, and the sporangium opens by a diagonal line.

Local Distribution
Confined to the most humid climatic portion of the coast from the Aleutian Chain, Alaska, southward to northern Vancouver Island and the adjacent mainland of the coastal areas in British Columbia.

World Distribution
Western North America: Alaska and British Columbia; East Asia: Japan, Borneo, and the Himalayan Mountains.

Distinguishing Characteristics
The small erect plants that form bright green turfs and have a colorless slender rhizomatous system and the individual leaves radially arranged on the shoots are very distinctive characteristics.

Similar Genera
The genus *Blepharostoma* is often similar in size and color, but the leaves are divided into narrow lobes; the plants also lack a colorless rhizomatous system. Perianths are also frequent in *Blepharostoma*, absent in *Takakia*.

Comments
This genus is so distinctive that it is sometimes placed into its own isolated group of plants completely independent of the rest of the bryophytes. In significant features, however, it is clearly a moss. The gametophyte has many features common to mosses, but many features are unique, in the bryophytes, to this genus. The sporophyte is decidedly of a moss, as is the calyptra. It could be dubbed the "puzzle plant" because of its baffling relationships. The Japanese name *nanja-monja-goke* (literally "impossible moss") reflects this bafflement. The genus has been included here because the plant so resembles a leafy liverwort.

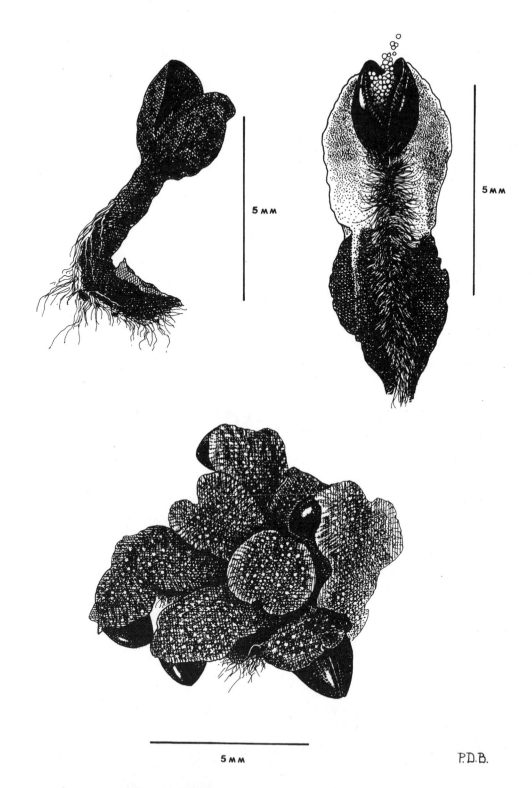

Targionia hypophylla from Californian specimen.
Upper right shows a thallus, with sporangium, viewed from beneath.

Targionia L.

Name
In honor of G. Targioni-Tozzeti, an eighteenth-century Florentine botanist and artist.

Species
Probably two species in the region.

Habit
Powdery-green thalli with nearly black scales on the undersurface, the sporangium sheathed by a glossy, purple black keeled envelope at the apex of the dorsal surface of the thallus. The thalli, when dry, have incurved margins and form black strands creeping on the substratum.

Habitat
Somewhat shaded fine-textured mineral soil on outcrop ledges in the northern part of its range, and in California abundant on shaded banks of the oak woodland and chaparral.

Reproduction
Sporophytes frequent in spring.

Local Distribution
From the southwestern portion of British Columbia mainly near the coast, and southward to southern California, where it is often abundant.

World Distribution
Widely distributed in warm temperate to tropical parts of the world.

Distinguishing Characteristics
The powdery green thalli with dark purple ventral scales and the immersed sporophyte on the ventral surface at the thallus apex are usually enough to separate this genus. Fortunately, sporophytes are usually present.

Similar Genera
When bearing sporophytes, no other local genus can be mistaken for *Targionia*. Sterile specimens of *Mannia* are similar in color, but the thallus tends to be very elongate and strongly incurved to wormlike when dry, and never so in *Targionia*.

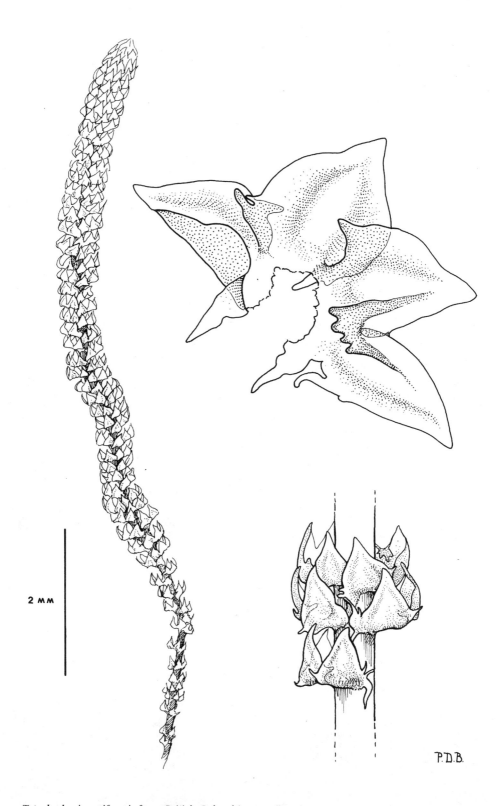

Tetralophozia setiformis from British Columbian specimen.

Tetralophozia (Schust.) Schljak.

Name
Means "four-pointed," referring to the neatly four-pointed leaves.

Species
Two species in the region.

Habit
Forming dark rusty brown to orange brown turfs of slender shoots in which the leaves are neatly and regularly four-lobed with reflexed margins, and the lobes bend upward parallel to the stem. Plants can be less than 1 mm to more than 2 mm in diameter; shoot length varies from 1 to 10 cm.

Habitat
T. filiformis occurs mainly on somewhat shaded cliffs near sea level, while *T. setiformis* is commonly among boulders of screes and talus slopes and in tundra habitats.

Reproduction
Sporophytes unknown in local material; gemmae noted rarely in *T. setiformis*. Plants are brittle when dry, and fragments probably serve in propagation.

Local Distribution
T. filiformis is confined to near the humid climatic areas of the coast in southwestern Alaska and British Columbia. *T. setiformis* is widely distributed through Alaska and British Columbia, but most frequent in the north.

World Distribution
The genus is predominantly in cooler climates of the Northern Hemisphere, where it is widespread.

Distinguishing Characteristics
The rusty brown plants with very regularly four-lobed leaves that arch upward abruptly to make the leafy shoots appear to have many rows of leaves make this genus very distinctive.

Similar Genera
Chandonanthus is extremely similar, but the leaves are three-lobed with many regular marginal teeth, and the plants tend to be yellowish to golden rather than rusty brown. *Herbertus* is similar in size and form, but its leaves are bilobed. *Anastrophyllum* is sometimes similar, but its leaves are also bilobed.

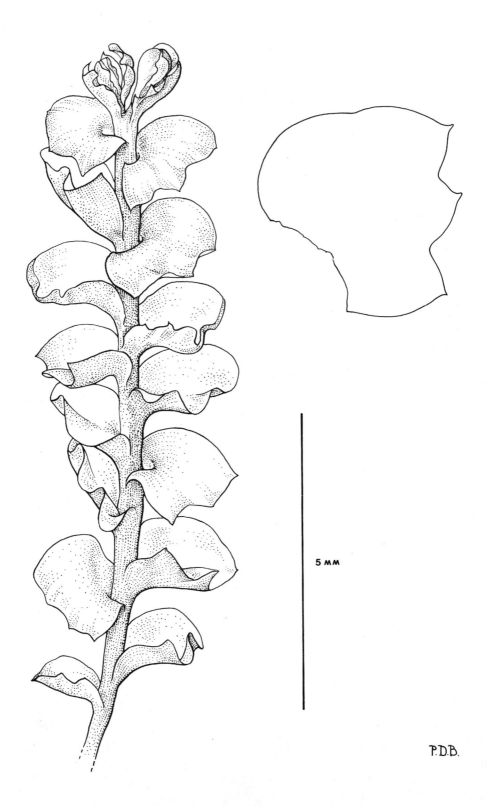

Tritomaria quinquedentata from British Columbian specimen.

Tritomaria Schiffn. ex Loeske

Name
Means "thrice cut," in reference to the three-lobed leaves.

Species
Six species in the region.

Habit
Pale to dark green turfs of erect to suberect or reclining unbranched or little-branched shoots. Leaves usually have two equal lobes and one markedly smaller lobe.

Habitat
Mainly in humid sites, on logs, rocks, humus, and earth, occasionally epiphytic, from sea level to alpine elevations.

Reproduction
Most species produce gemmae on the lobe tips of leaves near shoot apices. Gemmae are reddish or red brown in many cases, but can be yellowish green. Sporophytes are infrequent and mature in late summer.

Local Distribution
Probably widespread in Alaska southward to Oregon, especially at higher elevations.

World Distribution
Widely distributed in arctic to temperate latitudes of the Northern Hemisphere.

Distinguishing Characteristics
The asymmetrically three-lobed leaves with the lobes often broadly triangular and confined to the upper one third of the leaf length and the absence of underleaves are useful distinguishing characteristics. Most of our species have one lobe conspicuously smaller than the other two.

Similar Genera
Some species of *Lophozia* resemble *Tritomaria*, but leaves are usually two-lobed or equally three- to four-lobed. In *Barbilophozia*, too, leaves tend to be four-lobed rather than three-lobed, and underleaves are present. In *Chandonanthus*, also with three-lobed leaves, the lobes bear distinctive teeth, lacking in *Tritomaria*.

Glossary

acute having a sharp but not extended point, often in reference to leaf apex

alkaline usually in reference to a substratum in which lime dominates

alpine the elevation above the limit of trees on mountains

annual surviving for one year

appendage referring to an attachment with a pinched base, at the tip of a scale on the undersurface of a thallose liverwort

antheridium (antheridia) the male sex organ of a bryophyte, usually consisting of a stalk and a spherical or elongate sac containing many sperms enclosed by a sterile jacket

archegonium (archegonia) the female sex organ of a bryophyte, usually consisting of a flask-shaped portion containing a single egg when mature

asexual reproducing without the fusion of sperm and egg, often termed "vegetative reproduction" in plants

asymmetric indicating orientation that is off center, or shape that has the portion on one side of the midline smaller than the other

bifid forked into two equal parts

bilateral symmetry an object divided into halves so that one side essentially mirrors the other

bilobed two lobes or divisions

blade the flattened portion of a leaf

bog (adj.: **boggy**) a vegetation type in which the water table is near the surface, the water is predominantly added through direct precipitation on the surface, and the predominant plant is peat moss (*Sphagnum*), growing on a peat substratum

bryophyte the group of land plants in which the sporophyte produces a single sporangium and is dependent on the perennial gametophyte for growth

calyptra the protective sleeve of tissue, composed of cells associated with the archegonium, that protects the embryonic sporophyte

ciliate with hairlike structure

complex pore a pore in a thallus that is bounded by a minute barrel several cells in height

cosmopolitan showing a worldwide geographic distribution, i.e., found almost everywhere

decurrent refers to leaf attachment, where the base of the leaf extends lengthwise or obliquely downward on one or both sides of the leaf margin

dichotomous equally forked to form a Y

disjunctive showing a conspicuously interrupted geographic distribution, usually related to climatic or historic factors

dorsal on the back; in liverworts when applied to the gametophyte, referring to the upper surface of the reclining plant

elater a cell within the sporangium in which the walls usually have coiled thickenings; the elaters coil and uncoil abruptly in response to moisture changes, throwing spores from the opening sporangium

elliptic an outline in which an elongate organ is expanded in the mid-portion and gradually tapers to the two ends

elongate longer than wide

emergent usually in reference to sporangia that extend on a seta just beyond the associated enveloping structures

endemic restricted to a confined geographic distribution

entire in reference to leaf or thallus margins that lack any teeth

epilithic growing on rock

epiphyllous growing on leaf surfaces

epiphyte growing on another plant, using the other plant as a surface on which to perch

flagellate whiplike, usually referring to very slender shoots with distant or no leaves

flagellum a whiplike structure, usually a branch of conspicuously slenderer dimensions than the main shoot; in the sperm the whiplike structure that gives the sperm motility

foot the base of the sporophyte that penetrates the gametophyte to attach the sporophyte firmly, and through which nutrients are transferred

gametophyte the plant that produces the sex organs

gemma (gemmae) a specialized vegetative reproductive structure produced on the leafy or thallose plant, very loosely attached when mature

genus (genera) a term used to classify a rank of organisms that shows a constellation of features unique to that group and not identical to any other group of organisms of the same rank

humic of decomposed plant remains forming a strongly organic soil

humid of moderate moisture in the substratum, or high water vapor in the air around an organism

humus organic soil

imbricate overlapping, often used to describe leaves that are closely compressed to the stem

immersed referring to sporangia that are sheathed by an enveloping structure

incubous referring to leaf arrangement and orientation in leafy liverworts which, viewed on the dorsal surface, show the upper margin of a leaf visible and overlapping the lower part of the leaf above

involucre a sleeve or flap of thallus that encloses the archegonia and the developing sporophyte

lacerate referring to elongate narrow teeth, making a margin appear torn into elongate teeth

lamina equivalent to blade, the flattened portion of the leaf

lateral referring to the sides, rather than to the upper or under surfaces

linear in reference to an organ that is strongly elongated with parallel sides

lobe in reference to leaves, denoting that the leaf blade is divided longitudinally into two or more parts, each of which is a lobe; in a thallus, indicating that the thallus is similarly divided longitudinally, or has lateral flaps on the margin

lobule in lobed leaves, referring to the smaller of two lobes

longitudinal lengthwise

marsupium a pouchlike part of the ventral surface of the stem of a leafy liverwort that contains archegonia, and therefore the young sporophyte

meiosis nuclear division, in bryophytes always in the sporangial cells that produce spores, in which the chromosome number is reduced to one half that of the sporophyte

midrib an abruptly thickened central line in thallose liverworts

oblique in reference to leaf attachment, indicating that the line of attachment is slanting rather than at right angles to the length of the stem

obtuse blunt, often in reference to the leaf, shoot, or thallus apex

oceanic near the ocean, often indicating more humid climate than areas away from the coast

operculum a lid of a sporangium, present in a few thallose liverworts and most mosses

organ a structure in a plant that has a particular function (a leaf is an organ)

ovoid of the general shape of a bird's egg, thus tapered on one end

papilla a small wartlike projection

peat a brown fibrous turf that is formed mainly of partly decomposed vegetable matter, especially of peat moss

perennial living more than a year

perianth a sleevelike structure formed of fused leaves enclosing the female sex organs and thus the developing sporophyte of most leafy liverworts

perigonia protective leaves that enclose the antheridia

perigynium the flaps or scales that enclose the female sex organs of thallose liverworts, equivalent to the involucre

pinnate referring to a regular feather-like branching pattern in which branches diverge regularly in one plane on two opposite sides of a central axis

pore in thallose liverworts, referring to holes in the dorsal surface that allow gas exchange to the photosynthetic air chambers beneath

radial symmetry denoting that an object can be divided in many cases into many portions each resembling the other, thus the parts of that object radiating out from the center

receptacle in thallose liverworts, referring to a specialized perpendicular branch arising from the thallus that holds the male sex organs or the female sex organs and ultimately the sporophytes

recurved curved downward, usually on the margins, toward the ventral side

reflexed strongly curled downward and under on the margins, toward the ventral side

rhizoid unicellular hairs that attach liverworts and hornworts to their substratum and are important in the water economy of the plant

rhizome a multicellular reclining part of the stem, usually leafless

scale in liverworts, referring to leaflike, but often not green, flattened structures on the stem or thallus

seta the stalk of a sporophyte

sexual reproduction reproduction that requires the sperm to fertilize the egg, ultimately generating a sporophyte that produces spores

siliceous referring to a substratum in which quartz sand or its derivatives predominate, making an acidic surface or medium

simple pore a pore in a thallus in which the bounding cell layer is a single cell in thickness

sinuous bending in and out like a series of rounded waves

sinus the indentation between lobes

species a term used to classify a rank of organisms that belong to a genus, and each differing from the other by a suite of characters

sporangium the multicellular organ in bryophytes that produces and contains spores

spore the final reproductive structure that is produced by a sporangium and allows for dissemination of the plant as well as contributes to its evolutionary potential

sporophyte   the entire organ system that results from sexual reproduction

stem calyptra   a sheath of gametophytic tissue derived from both the archegonium and considerable tissue of the stem beneath it

subalpine   the slopes on mountains just below the limit of trees

suberect   nearly erect

substratum   the medium upon which plants are affixed

succubous   referring to leaf arrangement and orientation in leafy liverworts in which the upper margin of a leaf pushes under the lower margin of the leaf above it, thus the upper margin is not usually visible when viewed toward the dorsal surface

taxon (taxa)   a term referring to any named taxonomic group; each genus and species represents a different taxon

terrestrial   growing on the ground

thallus (thalli)   a flattened straplike plant body in which leaves and stems are usually not discernible

transverse   across the length of an object, when referring to leaf attachment, indicating that the attachment is a right angle, not an oblique angle, to the length of the stem

tuber   in thallose liverworts and hornworts denoting a multicellular vegetative reproductive organ that contains an apical cell sheathed by dead cells, and allows the gametophyte to survive unfavorable conditions for a limited period

turf   a growth form in which plant shoots are parallel to each other and erect on the substratum, forming a tight mat of shoots

underleaf   in leafy liverworts, the leaf on the ventral surface of the stem, flanked by the lateral leaves

unlobed   referring to leaves or thalli that lack any divisions in the outline

vegetative reproduction   reproduction involving any means of fragmentation of the living plant, each fragment of which has the potential to grow into an independent plant

ventral   the surface of the plant next to the substratum

zygote   the first cell of the sporophyte, resulting from the fertilization of the egg by the sperm

Checklist

A Checklist of Hepaticae and Anthrocerotae in Pacific North America

The following list includes all hepatic and hornwort species reported from the region covered in the book. *Takakia* is also included, although it is a moss. The following explains why the authors of species are included as part of the scientific name.

When a new genus or species is described, it is provided with a description in Latin and given a name in Latin or latinized Greek. The author of the genus is supplied so that the literature in which the original description was presented can be located for reference. The genus *Acrobolbus* Nees was first recognized by Nees, who published its description in 1844. The species *Acrobolbus ciliatus* (Mitt.) Schiffn. was first described by Mitten in 1861, but he thought it belonged to the genus *Gymnanthe* Tayl. In 1893, Schiffner transferred the species to the genus *Acrobolbus*; therefore his name is cited to indicate who made this decision.

Acrobolbus ciliatus (Mitt.) Schiffn.
Anastrepta orcadensis (Hook.) Schiffn.
Anastrophyllum assimile (Mitt.) Steph.
Anastrophyllum cavifolium (Buch & S. Arn.) Lammes
Anastrophyllum donnianum (Hook.) Spruce
Anastrophyllum hellerianum (Nees) Schust.
Anastrophyllum joergensenii Schiffn.
Anastrophyllum michauxii (Web.) Buch *ex* Evans
Anastrophyllum minutum (Schreb.) Schust. (var. *minutum* and var. *grandis* [Gott. *ex* Lindb.] Schust.)
Anastrophyllum saxicolum (Schrad.) Schust.
Anastrophyllum spenoloboides Schust.
Aneura pinguis (L.) Dum.

Anthelia julacea (L.) Dum.
Anthelia juratzkana (Limpr.) Trev.
Anthoceros bulbiculosus Brotero.
Anthoceros carolinianus Michx.
Anthoceros hallii Aust.
Anthoceros oreganus Aust.
Anthoceros pearsonii M. A. Howe
Apometzgeria pubescens (Schrank) Kuwah.
Apotreubia hortonae Schust. & Konst.
Arnellia fennica (Gott.) Lindb.
Ascidiota blepharophylla Massal.
Aspiromitus fusiformis (Aust.) Schust.
Aspiromitus punctatus (L.) Schljakov
Asterella bolanderi (Aust.) Underw.
Asterella californica (Hampe) Underw.
Asterella gracilis (Web.) Underw.
Asterella lindenbergiana (Corda) Lindb.
Asterella palmeri (Aust.) Underw.
Asterella saccata (Wahlenb.) Evans
Athalamia hyalina (Sommerf.) Hatt.

Barbilophozia atlantica (Kaal.) K. Muell.
Barbilophozia attenuata (Mart.) Loeske
Barbilophozia barbata (Schmid. *ex* Schreb.) Loeske
Barbilophozia binsteadii (Kaal.) Loeske
Barbilophozia floerkei (Web. & Mohr) Loeske
Barbilophozia hatcheri (Evans) Loeske
Barbilophozia hyperborea (Schust.) R. Stotl. & B. Stotl. *ex* Potemk.
Barbilophozia kunzeana (Hueb.) Gams
Barbilophozia lycopodioides (Wallr.) Loeske
Barbilophozia quadriloba (Lindb.) Loeske
Bazzania denudata (Torrey *ex* Gott. *et al.*) Trev.
Bazzania pearsonii Steph.
Bazzania tricrenata (Wahlenb.) Lindb.
Bazzania trilobata (L.) S. Gray
Blasia pusilla L.

Blepharostoma arachnoideum M. A. Howe
Blepharostoma trichophyllum (L.) Dum. (ssp. *trichophyllum* and ssp. *brevirete* (Bryhn. & Kaal.) Schust.
Bucegia romanica Radian

Calycularia crispula Mitt.
Calycularia laxa Lindb. & Arnell
Calypogeia azurea Stotler & Crotz
Calypogeia fissa (L.) Raddi
Calypogeia integristipula Steph.
Calypogeia muelleriana (Schiffn.) K. Muell.
Calypogeia neesiana (Mass. & Carest.) K. Muell.
Calypogeia sphagnicola (H. Arnell & J. Perss.) Warnst. & Loeske
Calypogeia suecica (H. Arnell & J. Perss.) K. Muell.
Cephalozia affinis Lindb. ex Steph.
Cephalozia bicuspidata (L.) Dum. (ssp. *bicuspidata*, ssp. *ambigua* [Maas.] Schust. and ssp. *lammersiana* [Hueb.] Schust.)
Cephalozia catenulata (Hueb.) Lindb.
Cephalozia connivens (Dicks.) Lindb.
Cephalozia leucantha Spruce
Cephalozia lunulifolia (Dum.) Dum.
Cephalozia macounii (Aust.) Aust.
Cephalozia pachycaulis Schust.
Cephalozia pleniceps (Aust.) Lindb.
Cephaloziella arctica Bryhn & Douin
Cephaloziella arctogena (Schust.) Konst.
Cephaloziella aspericaulis Joerg.
Cephaloziella brinkmanii Domin
Cephaloziella divaricata (Sm.) Schiffn. (var. *divaricata* and var. *scabra* M. A. Howe)
Cephaloziella elachista (Jack) Schiffn.
Cephaloziella elegans (Heeg) Schiffn.
Cephaloziella grimsulana (Jack. *ex* Gott. & Rabenh.) Lacout
Cephaloziella hampeana (Nees) Schiffn.
Cephaloziella integerrima (Lindb.) Warnst.
Cephaloziella phyllacantha (Mass. & Carest.) K. Muell.
Cephaloziella rubella (Nees) Warnst.
Cephaloziella spinigera (Lindb.) Joerg.

Cephaloziella stellulifera (Tayl. *ex* Spruce) Schiffn.
Cephaloziella subdentata Warnst.
Cephaloziella turneri (Hook.) K. Muell.
Cephaloziella uncinata Schust.
Chandonanthus hirtellus (Web.) Mitt.
Chiloscyphus gemmiparus (Evans) Dum.
Chiloscyphus pallescens (Ehrh. ex Hoffm.) Dum.
Chiloscyphus polyanthos (L.) Corda (var. *polyanthos* and var. *rivularis* [Schrad.] Nees)
Cladopodiella fluitans (Nees) Joerg.
Cololejeunea macounii (Spruce *ex* Underw.) Evans
Conocephalum conicum (L.) Lindb.
Cryptocolea imbricata Schust.
Cryptomitrium tenerum (Hook.) Aust.

Dendrobazzania griffithiana (Steph.) Schust. & Schof.
Diplophyllum albicans (L.) Dum.
Diplophyllum imbricatum (M. A. Howe) K. Muell.
Diplophyllum microdontum (Mitt.) Buch
Diplophyllum obtusifolium (Hook.) Dum.
Diplophyllum plicatum Lindb.
Diplophyllum taxifolium (Wahlenb.) Dum. (var. *taxifolium* and var. *macrostictum* Buch.)
Douinia ovata (Dicks.) Buch.

Eremonotus myriocarpus (Carring.) Pears.

Fossombronia alaskana Steere & Inoue
Fossombronia foveolata Lindb.
Fossombronia hispidissima Steph.
Fossombronia longiseta Aust.
Frullania bolanderi Aust.
Frullania californica (Aust.) Evans
Frullania catalinae Evans
Frullania chilcootenis Steph.
Frullania eboracensis Gottsche
Frullania franciscana M. A. Howe
Frullania hattoriana J. D. Godfr. & G. A. Godfr.
Frullania jackii Gottsche
Frullania nisquallensis Sull.

Geocalyx graveolens (Schrad.) Nees
Geothallus tuberosus Campb.
Gymnocolea acutiloba (Schiffn.) K. Muell.
Gymnocolea fascinifera Potemk.
Gymnocolea inflata (Huds.) Dum.
Gymnomitrion apiculatum (Schiffn.) K. Muell.
Gymnomitrion concinnatum (Lightf.) Corda
Gymnomitrion corallioides Nees
Gymnomitrion mycrophorum Schust.
Gymnomitrion obtusum (Lindb.) Pears.
Gymnomitrion pacificum Grolle
Gyrothyra underwoodiana M. A. Howe

Haplomitrium hookeri (Sm.) Nees
Harpanthus flotovianus (Nees) Nees
Herbertus aduncus (Dicks) S. Gray
Herbertus haidensis Schof.
Herbertus sakuraii Steph. (ssp. *sakuraii* and ssp. *arcticus* Inoue & Steere)
Herbertus sendtneri (Nees) Lindb.
Hygrobiella laxifolia (Hood.) Spruce

Jamesoniella autumnalis (D.C.) Steph.
Jamesoniella nipponica Hatt.
Jungermannia atrovirens Dum.
Jungermannia borealis Damsholt & Vaña
Jungermannia caespitica Lindb.
Jungermannia confertissima Nees
Jungermannia evansii Vaña
Jungermannia exsertifolia Steph. (ssp. *cordifolia* [Dum.] Vaña)
Jungermannia fusiformis (Steph.) Steph.
Jungermannia hyalina Lyell
Jungermannia leiantha Grolle
Jungermannia obovata Nees
Jungermannia polaris Lindb.
Jungermannia pumila With.
Jungermannia rubra Gott. *ex* Underw.
Jungermannia schusterana J. D. Godfr. & G. A. Godfr.
Jungermannia sphaerocarpa Hook

Kurzia pauciflora (Dicks.) Grolle
Kurzia sylvatica (Evans) Grolle
Kurzia trichoclados (K. Muell.) Grolle

Lejeunea alaskana (Schust. & Steere) Inoue & Steere
Lepidozia filamentosa (Lehm. & Lindenb.) Lindenb.
Lepidozia reptans (L.) Dum.
Lepidozia sandvicensis Lindenb.
Lophocolea cuspidata (Nees) Limpr.
Lophocolea heterophylla (Schrad.) Dum.
Lophocolea minor Nees
Lophozia alboviridis Schust.
Lophozia ascendens (Warnst.) Schust.
Lophozia badensis (Gottsche *ex* Gottsche & Rabenh.) Schiffn.
Lophozia bantriensis (Hook.) Steph.
Lophozia collaris (Nees) Dum.
Lophozia debiliformis Schust. & Damsch.
Lophozia decolorans (Limpr.) Steph.
Lophozia ehrhartiana (Web.) Inoue & Steere
Lophozia elongata Steph.
Lophozia excisa (Dicks.) Dum.
Lophozia gillmanii (Aust.) Schust.
Lophozia groenlandica (Nees *in* G. L. & N.) Macoun
Lophozia guttulata (Lindb. & H. Arnell) Evans
Lophozia heterocolpos (Thed. *ex* Hartm.) M. A. Howe
Lophozia holmenianum Inoue & Steere
Lophozia hyperarctica Schust.
Lophozia incisa (Schrad.) Dum.
Lophozia jurensis Meyl. *ex* K. Muell.
Lophozia laxa (Lindb.) Grolle
Lophozia longidens (Lindb.) Macoun
Lophozia obtusa (Lindb.) Evans
Lophozia opacifolia Culm.
Lophozia pellucida Schust.
Lophozia perssonii Buch & S. Arn.
Lophozia polaris Schust. & Damsholt
Lophozia rutheana (Limpr.) M. A. Howe
Lophozia sudetica (Hueb.) Grolle
Lophozia ventricosa (Dicks.) Dum. (var. *ventricosa* var. *longiflora* [Nees] Macoun, and var. *silvicola* [Buch] Jones)
Lophozia wenzelii (Nees) Steph.
Lunularia cruciata (L.) Dum.

Mannia californica (Gottsche) Wheeler
Mannia fragrans (Balbis) Frye & Clark
Mannia pilosa (Hornem.) Frye & Clark
Marchantia polymorpha L.
Marsupella alpina (Mass. & Carest.) H. Bern.
Marsupella aquatica (Lindenb.) Schiffn
Marsupella arctica (Berggr.) Bryhn. & Kaal.
Marsupella boeckii (Aust.) Kaal. (var. *stableri* [Spruce] Schust.)
Marsupella bolanderi (Aust.) Underw.
Marsupella brevissima (Dum.) Grolle
Marsupella commutata (Limpr.) H. Bern.
Marsupella condensata (Aongstr.) Lindb.
Marsupella emarginata (Ehrh.) Dum.
Marsupella revoluta (Nees) Dum.
Marsupella sparsifolia (Lindb.) Dum.
Marsupella sphacelata (Gieseke) Dum.
Marsupella spiniloba Schust. & Damsch.
Marsupella sprucei (Limpr.) H. Bern.
Marsupella tubulosa Steph.
Mastigophora woodsii (Hook.) Nees
Mesoptychia sahlbergii (Lindb. & Arn.) Evans
Metacalypogeia cordifolia (Steph.) Inoue
Metacalypogeia schusterana Hattori & Mizut.
Metzgeria conjugata Lindb.
Metzgeria leptoneura Spruce
Metzgeria temperata Kuwah.
Moerckia blyttii (Moerck) Brockm.
Moerckia hibernica (Hook.) Gott.
Mylia anomala (Hook.) S. Gray
Mylia taylorii (Hook.) S. Gray

Nardia breidleri (Limpr.) Lindb.
Nardia compressa (Hook.) S. Gray
Nardia geoscyphus (De Not.) Lindb.
Nardia insecta Lindb.
Nardia japonica Steph.
Nardia scalaris S. Gray

Odontoschisma denudatum (Nees *ex* Mart.) Dum.
Odontoschisma elongatum (Lindb.) Evans
Odontoschisma macounii (Aust.) Underw.
Odontoschisma sphagni (Dicks.) Dum.

Pallavicinia lyellii (Hook.) S. F. Gray
Pellia endiviifolia (Dicks.) Dum.
Pellia epiphylla (L.) Corda
Pellia neesiana (Gott.) Limpr.
Peltolepis quadrata (Saut.) K. Muell
Plagiochila arctica Bryhn. & Kaal.
Plagiochila poeltii Grolle
Plagiochila porelloides (Torrey *ex* Nees) Lindenb.
Plagiochila schofieldiana H. Inoue
Plagiochila semidecurrens Lehm. & Lindenb. (var. *semidecurrens* and var. *alaskana* [Evans] H. Inoue)
Pleurocladula albescens (Hook.) Grolle
Pleurozia purpurea Lindb.
Porella bolanderi (Aust.) Pears.
Porella cordaeana (Hueb.) Moore
Porella navicularis (Lehm. & Lindenb.) Lindenb.
Porella platyphylla (L.) Pfeiff.
Porella roellii Steph.
Porella vernicosa Lindb. (var. *fauriei* [Steph.] M. Hara)
Preissia quadrata (Scop.) Nees
Pseudolepicolea fryei (Perss.) Grolle & Ando
Ptilidium californicum (Aust.) Underw.
Ptilidium ciliare (L.) Hampe
Ptilidium pulcherrimum (G. Web.) Hampe

Radula auriculata Steph.
Radula bolanderi Gott.
Radula brunnea Steph.
Radula complanata (L.) Dum.
Radula obtusiloba Steph. (ssp. *polyclada* [Evans] Hatt.)
Radula prolifera H. Arnell
Reboulia hemisphaerica (L.) Raddi
Riccardia chamedryfolia (With.) Grolle
Riccardia latifrons Lindb.
Riccardia multifida (L.) S. Gray
Riccardia palmata (Hedw.) Carruth.
Riccia austinii Steph.
Riccia beyrichiana Hampe *ex* Lehm.
Riccia californica Aust.
Riccia campbelliana M. A. Howe

Riccia canaliculata Hoffm.
Riccia cavernosa Hoffm.
Riccia crystallina L.
Riccia fluitans L.
Riccia frostii Aust.
Riccia glauca L.
Riccia lamellosa Raddi
Riccia nigrella D. C.
Riccia sorocarpa Bisch.
Riccia trichocarpa M. A. Howe
Riccia violacea M. A. Howe
Ricciocarpos natans (L.) Corda
Riella affinis M. A. Howe & Underw.
Riella americana M. A. Howe & Underw.

Sauteria alpina (Nees) Nees
Scapania americana K. Muell.
Scapania apiculata Spruce
Scapania bolanderi Aust.
Scapania brevicaulis Tayl.
Scapania compacta (A. Roth) Dum.
Scapania crassiretis Bryhn
Scapania curta (Mart.) Dum.
Scapania cuspiduligera (Nees) K. Muell.
Scapania degenii Schiffn. *ex* K. Muell. (var. *dubia* Schust.)
Scapania diplophylloides Amak. & Hatt.
Scapania glaucecephala (Tayl.) Aust.
Scapania granulifera Evans
Scapania gymnostomophila Kaal.
Scapania hians Steph. (ssp. *salishensis* J. D. Godfr. & G. A. Godfr.)
Scapania hollandiae Hong
Scapania hyperborea Joerg.
Scapania irrigua (Nees) Gott.
Scapania kaurinii Ryan
Scapania massalongoi K. Muell.
Scapania mucronata Buch.

Scapania obcordata (Berggr.) S. Arn.
Scapania obscura (H. Arnell & C. Jens.) Schiffn.
Scapania ornithopoides (With.) Waddell
Scapania paludicola Loeske & K. Muell.
Scapania paludosa (K. Muell.) K. Muell.
Scapania parvifolia Warnst.
Scapania perssonii Schust.
Scapania praetervisa Meylan
Scapania scandica (H. Arnell & Buch) Macv.
Scapania simmonsii Bryhn & Kaal.
Scapania spitzbergensis (Lindb.) K. Muell.
Scapania subalpina (Nees) Dum.
Scapania tundrae (H. Arn.) Buch
Scapania uliginosa (Sw. ex Lindenb.) Dum.
Scapania umbrosa (Schrad.) Dum.
Scapania undulata (L.) Dum. (var. *undulata* and var. *oakesii* [Aust.] Buch)
Scapania zemliae S. Arn.
Schofieldia monticola J. D. Godfr.
Sphaerocarpos cristatus M. A. Howe
Sphaerocarpos drewei Wigglesworth
Sphaerocarpos hians Haynes
Sphaerocarpos michelii Bell.
Sphaerocarpos texanus Aust.
Sphenolobopsis pearsonii (Spruce) Schust.

Takakia ceratophylla (Mitt.) Grolle
Takakia lepidozioides Hatt. & H. Inoue
Targionia hypophylla L.
Tetralophozia filiformis (Steph.) Lammes
Tetralophozia setiformis (Ehrh.) Schljakov
Tritomaria exsecta (Schrad.) Loeske
Tritomaria exsectiformis (Breidl.) Loeske
Tritomaria heterophylla Schust.
Tritomaria polita (Nees) Joerg.
Tritomaria quinquedentata (Huds.) Buch
Tritomaria scitula (Tayl.) Joerg.

References

The following publications, discussed briefly in the Introduction, are useful references for the region.

Clark, L., and T. C. Frye. "The Liverworts of the Northwest," *Publication of the Puget Sound Biological Station* 6 (1928): 1–193.

Davison, P. G. "Floristic and Phytogeographic Studies of the Hepatic Flora of the Aleutian Islands, Alaska." Ph.D. diss., University of Tennessee, Knoxville, 1994.

Godfrey, J. D. "The Hepaticae and Anthocerotae of Southwestern British Columbia." Ph.D. diss., University of British Columbia, 1977.

Howe, M. A. "Hepaticae and Anthocerotae of California," *Memoirs Torrey Botanical Club* 1 (1899): 1–208.

Macvicar, S. M. *The Student's Handbook of British Hepatics.* Eastbourne, England: V. V. Sumfield, 1926.

Paton, J. A. *The Liverwort Flora of the British Isles.* Colchester, England: Harley Books, 1999.

Sanborn, E. L. "Hepaticae and Anthocerotae of Western Oregon," *University of Oregon Plant Biology,* ser. 1, no. 1 (1929): 1–111.

Schofield, W. B. *Introduction to Bryology.* New York: Macmillan Company, 1985.

Schuster, R. M. *The Hepaticae and Anthocerotae of North America.* 4 vols. New York: Columbia University Press, 1966–92.

———*The Hepaticae and Anthocerotae of North America.* 2 vols. Chicago: Field Museum of Natural History, 1992.

Steere, W. C., and H. Inoue. "The Hepaticae of Arctic Alaska," *Journal of Hattori Botanical Laboratory* 44 (1978): 251–345.

Whittemore, A. L. "A Preliminary Checklist, with Keys, of California Liverworts and Hornworts." Manuscript, 1999.

Index

Acrobolbus ciliatus, 36-37
Ahti, T., 14
allergies, 22, 105
Anastrepta orcadensis, 38-39
Anastrophyllum assimile, 147
Anastrophyllum minutum, 40-41
Aneura pinguis, 42-43
Anomomarsupella, 147
Anthelia julacea, 44-45, 125
antheridium, 4
Anthoceros carolinianus, 46-47
Apometzgeria pubescens, 48-49
Apotreubia hortonae, 50-51
aquatic liverworts, 16
archegonium, 4
Arnell, H. W., 53
Arnellia fennica, 52-53
Ascidiota blepharophylla, 54-55
Aspiromitus punctatus, 56-57
Asterella bolanderi, 15
Asterella californica, 59
Asterella gracilis, 58-59
Asterella lindenbergiana, 59
Athalamia hyalina, 60-61

Barbilophozia barbata, 62-63
Barbilophozia floerkei, 22
Barker, M., 13
Bazzani, M., 65
Bazzania denudata, 64-65
Blasia pusilla, 66-67
Blepharostoma arachnoideum, 69, 179
Blepharostoma trichophyllum, 68-69
Bolander, H. N., 15
boulders: as habitat, 17
Brinkman, A. H., 14
bryophytes, 3
Bucegia romanica, 70-71

Calycularia crispula, 72-73, 157
Calycularia laxa, 73
Calypogeia muelleriana, 74-75
calyptra, 4
Cephalozia bicuspidata, 76-77
Cephaloziella brinkmanii, 14
Cephaloziella divaricata, 78-79
Cephaloziella phyllacantha, 22
Chandonanthus hirtellus, 80
Chiloscyphus polyanthos, 82-83
Christy, J., 15
Cladopodiella fluitans, 84-85
Clark, Lois, 15
cliffs, and liverworts, 17
collecting procedures, 7-12
collectors. *See* history of collectors
Cololejeunea macounii, 14, 86-87
color, 6
Conocephalum conicum, 88-89
Cooley, Grace, 13
Coville, J. V., 13
Cryptocolea imbricata, 90-91
Cryptomitrium tenerum, 92-93

Davison, P. G., 13
Dendrobazzania griffithiana, 94-95
determination, 20
Diplophyllum albicans, 96-97
distribution patterns, 18-19
Douin, C. J., 99
Douinia ovata, 98-99
Doyle, W. T., 15, 108

elaters, 4
epiphytes, 17
Eremonotus myriocarpus, 100-101
Evans, Alexander, 13
Eyerdam, W. J., 13

fertilization, 4
flagellum, 4
Flowers, S., 14
Fossombroni, V., 103
Fossombronia longiseta, 102-3, 135
Foster, A. B., 13
Foster, A. S., 15
Frullani, L., 105
Frullania bolanderi, 15
Frullania hattoriana, 14
Frullania nisquallensis, 22, 104-5
Frye, T. C., 13-15

gametophyte, 3
gemma cup, 140, 144
gemmae 4
Geocalyx graveolens, 22, 106-7
Geothallus tuberosus, 108-9
Gibbs, Sarah, 14
Godfrey, Geoffrey, 14-15
Godfrey, Judith, 14-15
Gymnocolea inflata, 110-11
Gymnomitrion obtusum, 112-13
Gyrothyra underwoodiana, 114-15

habitats, 16-18
Hämet-Ahti, L., 14
Haplomitrium hookeri, 116-17
Harpanthus flotovianus, 118-19
Harpel, J., 15
Hepaticae and Anthocerotae of North America (Schuster), 21
"Hepaticae and Anthocerotae of Southwestern British Columbia" (Godfrey), 21
Hepaticae and Anthocerotes of California (Howe), 20
Hepaticae and Anthocerotes of Western Oregon (Sanborn), 20
Herbert, T., 121
Herbertus aduncus, 120-21, 179
history of collectors, 13-15
  Alaska, 13
  British Columbia, 14
  California, 15
  Oregon, 15
  Washington, 14-15
Holmen, K., 13
Hong, W. S., 14-15
hornworts: structure, 5; compared to mosses, 7
Horton, D. H., 13-14
Howe, M. A., 15
Hygrobiella laxifolia, 122-23

Inoue, H., 13
Iwatsuki, Z., 13

Jamesoniella automnalis, 124-25
Jamieson, William, 125
Jungermann, L., 127
Jungermannia atrovirens, 22
Jungermannia exsertifolia, 83, 159
Jungermannia rubra, 126-27
Jungermannia schusterana, 14

Kearney, T. H., 13
Kingsman, C. C., 15
Konstantinova, N., 13
Kurz, W. S., 129
Kurzia pauciflora, 128-29

lateral leaf, 8
Latin names, 21
leaf form, 9
Lejeune, L. S., 131
Lejeunea alaskana, 130-31
Lepidozia reptans, 132-33
Lepidozia sandvicensis, 133
liverworts
  allergies to, 22, 105
  aquatic, 16
  color, 6
  compared to lichens, 3
  compared to mosses, 7
  habitats, 16
  leafy structure, 8
  odor, 22, 89
  and people, 21-22

pollution indicators, 22
structure 3-6
taste, 22
weeds, 22
wetland, 16-17
The Liverwort Flora of the British Isles (Paton), 20
The Liverworts of the Northwest (Clark and Frye) 14, 20
lobe, 8
lobule, 8
Lophocolea cuspidata, 134-35
Lophocolea heterophylla, 83
Lophocolea minor, 135
Lophozia, 136-39
Lophozia gillmanii, 136
Lophozia incisa, 137
Lophozia rutheana, 151
Lophozia ventricosa, 138
Lunularia cruciata, 140-41

Macoun, John, 14
Macoun, J. M., 13
Mannia fragrans, 142-43
Marchant, N., 145
Marchantia polymorpha, 144-45
Marsupella emarginata, 146-47
marsupium, 8, 75, 106
Mastigophora woodsii, 148-49
Menzies, A., 15
Mesoptychia sahlbergii, 151-52
Metacalypogeia schusterana, 152-53
Metzer, J. B., 155
Metzgeria conjugata, 154-55
Metzgeria leptoneura, 155
Metzgeria temperata, 155
Miller, H. A., 13-15
Moerck, A., 157
Moerckia blyttii, 156-97
Moerckia hibernica, 157
Murray, B., 13
Mylia anomala, 159
Mylia taylorii, 125, 158-59

nanja-monja-goke, 205
Nardi, S., 161
Nardia compressa, 159
Nardia scalaris, 160-61
Norris, D. H., 15
Nostoc, 47, 57, 67

Odontoschisma denudatum, 162-63
Odontoschisma elongatum, 163
Odontoschisma gibbsiae, 14
Odontoschisma macounii, 163
Odontoschisma sphagni, 163
odor, 22, 89

Pellia neesiana, 164-65
Pelli-Fabbroni, L., 165
Peltolepis quadrata, 166-67
perianth, 8
Persson, Herman, 13-14
Plagiochila arctica, 169
Plagiochila poeltii, 169
Plagiochila porelloides, 83, 168-69
Plagiochila schofieldiana, 169
Plagiochila semidecurrens, 169
Pleurocladula albescens, 170-71
Pleurozia purpurea, 172
pollution indicators, 22
Porella bolanderi, 15
Porella havicularis, 174-75
Porella roellii, 22
Porella vernicosa var. *fauriei*, 55, 175
Potemkin, A., 13
Preiss, B., 177
Preissia quadrata, 176-77
Pseudolepicolea fryei, 178-79
Ptilidium californicum, 180-81
Ptilidium ciliare, 55, 181
Ptilidium pulcherrimum, 181

Radula auriculata, 183
Radula bolanderi, 15
Radula brunnea, 105, 183
Radula complanata, 182-83
Reboulia hemisphaerica, 184-85

receptacle form, 11
Riccardi, V., 187
Riccardia latifrons, 186–87
Ricci, P. F., 189
Riccia cavernosa, 191
Riccia fluitans, 187, 189
Riccia sorocarpa, 188–89
Ricciocarpos natans, 190–91
Riella affinis, 192–93
Rigg, G. B., 13
Rothrock, J. T., 13
rotten logs, 17

Sanborn, E. L., 15
Sauter, A., 195
Sauteria alpina, 194–95
Sauteria berteroana, 195
"scale mosses," 3
Scapania bolanderi, 15, 196–97
Scapania hians ssp. *salishensis*, 14
Scapania undulata, 22
Schofield, W. B., 13–15, 199
Schofieldia monticola, 14, 198–99
Schuster, R. M. , 13–14
Scouler, J., 15
seasonality, 17–18
seta, 4
Setchell, William A., 13
Shacklette, H., 13
Sharp, A. J., 13
Smith, D., 13
Sphaerocarpos texanus, 200–201
Sphenolobopsis pearsonii, 79, 202–3
sporangium form, 10
spore dispersal, 4
sporophyte, 4

Steere, W. C., 13
Storage of specimens, 12
The Student's Handbook of British Hepatics (Macvicar), 20
stumps, 17
Sutcliffe, D. L., 15

Takaki, N., 205
Takakia ceratophylla, 205
Takakia lepidozioides, 14, 204–5
Talbot, S., 13
Talbot, S. Looman, 13
Targioni-Terzetti, G., 207
Targionia hypophylla, 206–7
taste, 22
Tetralophozia filiformis, 81, 209
Tetralophozia setiformis, 81, 208–9
thallus, 3
Trelease, William, 13
Tritomaria quinquedentata, 210
tubers, 5
tundra, 18

underleaf, 8

Wagner, D., 15, 92
Weber, W., 108
weeds, 22
wetland liverworts, 16–17
Whittemore, A. T., 15
Williams, R. S., 13
Worley, I., 13

Yurky, P., 15

zygote, 4